WIPO | PROOF
Trusted Digital Evidence

CERTIFICATE ID: WDTS-PC-00002732

The World Intellectual Property Organization hereby certifies that,
on Jan 18, 2022 09:50 UTC, Olivier HAKIZIMANA had possession
of the digital file described below.

Digital file name:	ENERGY MAKER MACHINE
Digital fingerprint of the file:	23b6f3b149a9856370ff7f2ea97b5a1253bf176d00f329ceb7863ef01cd6a3ce

This certificate was issued on Jan 18, 2022 10:16 UTC upon presentation of the digital file and its related WIPO PROOF token.

Serial number: -2486219417182106497602993580162687969469388551681 Policy: 1.3.6.1.4.1.48669.2.1.1
Timestamping Authority: O = World Intellectual Property Organization | OI = CH | L=462.076.735 | C = CH

WIPO | PROOF
Trusted Digital Evidence
Certified Copy

Daren TANG
Director General
World Intellectual Property Organization

***   ENGLISH   ***

Unedited second original manuscript

THE UNIVERSAL MUTATION VIA THE ENERGY MAKER MACHINE

''A PLANETARY TRANSMISSION WITH A COMMON COMPOUND PLANET WHEEL, A STATIONARY RING WHEEL, AN INPUT SUN WHEEL, AN OUTPUT SUN WHEEL AND A RESTORER WHEEL''

Highlights

The present technology depicts new discovered fundamental laws of physics; and how, despite their classical simplicity, they clarify the so-called dark gravity and violate almost all principles already known like ''the momentum and/or energy conservation principle''.

Abstract

This technology rises from an exploration of how nature manages for frictions to subsist along with the motion while they were generally supposed to annihilate.

An experiment consisting of wheels introduced into a dynamic system that combines two paradoxes of visible bodies in a rolling-sliding motion (The Aristotle's wheel paradox and the paradox of a circle that rolls on another stationary circle); and resulting in a constitution of a device that generates mechanical energy.

An apparatus characterized in that it comprises a **"planetary transmission with a common compound planet wheel"** having in principle five wheels indirectly connected by means of chains or belts with reference to **FIG.1a** and **FIG.1c,** and directly connected by means of gears with reference to **FIG.2a**:

- Input compound sun wheel **(4)**.
- Common compound planet wheels **(2)**.
- Stationary ring wheels **(3)**.
- Output compound sun wheel **(1)**.
- Restorer wheels **(5)**.

Those wheels are hold together thanks to a rotating carrier **(7)** of one or more common compound planets and a central fixed axle **(6)**.

That carrier does not play a role of a wheel, it is instead a dynamically orbital-fixed-pulley.

Surprisingly, it ended up clarifying a scientific topic considered as an enigma (the so-called dark gravity) if not a taboo (the both creation and/or loss of energy from the visible to invisible, vice versa) since it is about violating "the momentum and/or energy conservation principle".

As a matter of fact, within any dynamically orbital system, the visible whole is created and lost within the invisible; it is a mutation (not just a simple transformation or a simple transfer of physical quantities) of which only the trigger of the movement decides its beginning, its magnitude and its end.

Keywords:

Orbital-friction
Mutational-factor
No-local-hidden-variable
Information-energy-equivalence
Universal-thinking and universal-certainty

Contents

Scientific writing	THE UNIVERSAL MUTATION [Micro-Invisible-Macro]
INTRODUCTION	**1. Frictions**
	A. The omnipresence of frictions
HYPOTHESIS	**2. The phenomenon of mutational frictions**
	2.1. Paradoxes of visible body in motion
OBSERVATION	2.2. Analysis of simple machines
EXPERIMENTATION	2.3. Experiment of double orbital friction
RESULTS	2.3.1. HAKIZIMANA's paradox
INTERPRETATION	2.3.2. The INGOMA
	2.3.3. The HAKIZA
	A Mutated physical quantity
	B.1 Mutational factorization
	B.2 Mutational invariance
	B.3 Mutational singularity
PROTOTYPE	C. The HAKIZIMANA's prototype machine
CONCLUSION	**3. The Introduction to the theory of everything**
	A. The natural HAKIZA
	3.1. The Theory of everything via Mutational factorization
	3.2. Universal thinking via mutational invariance
	3.3. Universal Certainty via Mutational singularity

INTRODUCTION

THE UNIVERSAL MUTATION [Micro-Invisible-Macro]

1. Friction

A. The omnipresence of frictions

New technologies seek to maximize the efficiency in energy transformation regardless of its form: ''Tribology and Nano-tribology to study and reduce or if possible make zero friction (supra-friction)''.

Fortunately and/or unfortunately, these technologies only succeed in making frictions negligible and this on a very small scale compared to the immensity of the cosmos; since a visible universe does not contain absolute emptiness, nature unceasingly causes frictions.
And they manifest themselves everywhere thanks to the existence of a continuous movement of the visible whole with respect to both itself and the invisible (respectively at the macroscopic and microscopic scale) by:

- A paradoxically methodical-systematic and continuous fall, for stars.
- A paradoxical impediment to achieving absolute immobility, for elementary particles.

And unlike many Scientists who, on our scale, have not been able to reproduce perpetual motion, instead of contradicting nature by assuming it to be impossible; we rather prefer to answer the below counterintuitive question.

From the exact-sciences point of view:
"how does nature manage for frictions to subsist along with the motion of the visible whole while they were supposed to annihilate"?

Equivalent from social-sciences perspective to:
"why something rather than nothing"?

Through a thought experiment, it turns out that there are three theoretically conceptual scenarios to understand before remedying the above cited enigma:

- **What happens to the visible and/or observable universe without frictions?**

→ Knowing that within the visible and/or observable universe, the visible whole is always in motion within a non-absolute emptiness;
We always get an incomplete idea that frictions have only as main property: ''the effect of systematically thwarting or preventing deformations and relative motions''.

At the end of the day, we neglect that there can never be any fundamental interaction (gravitational, electromagnetic, weak and strong nuclear) without there having been the slightest friction.

And that it would be insane to speak of a visible and/or observable universe without frictions; since frictions simultaneously and unceasingly allow and prevent motion at any point of the visible whole...!

- **What connects the visible and/or observable universe to the invisible?**

→ Simply by the logic of the cosmic essence of the term ''invisible (whose actual scientific approach is the so-called dark matter and/or dark energy)'', the invisible must be beyond and everywhere within the visible.

And the visible universe being always in motion, there is no objection to theoretically considering admitting *that the invisible and the visible are linked by a certain phenomenon of frictions obeying different laws than those already known.*

- **Would the creation and/or loss of only a very small portion of energy from the visible and/or observable universe be caused constantly and simultaneously by the same phenomenon?**

What a poorly worded question! Almost all scientists would say.

- This question would be foolish, given that the internal energy is a property of the state of a single system, therefore this system in question can be understood without knowing how the energy got there. Hence scientists are convinced that within the visible universe:

« *Nothing created, nothing lost; it is just a transformation* »!... ???

Unfortunately, this would probably only be valid within a visible universe considered as a closed system.
A system without any other internal, external and undetectable energy (commonly called dark energy preventing any attempt to isolate a system).
Nevertheless, this is far from being the case, which thus reinforces **GÖDEL**'s principles of incompleteness while making the principles of thermodynamics and dynamics unusable beyond the visible.

Fortunately, assuming that this dark energy is both beyond and everywhere within the visible and/or observable universe; *it is to admit that the notions of energy and singularity are incomplete*.

And to better address them, it would be essential to place oneself theoretically in a randomly invisible referential frame whose proprieties do not matter, according to a simple logic of arithmetic:
''It is logical to be in the invisible without being in the visible, while it is illogical to be in the visible without being in the invisible, since the visible is part of the invisible''.

In fact, all the dimensions and all possible referential frames of the visible whole are respectively miniaturized into a universal dimension and referential frame called *singularity*.

- Finally, it would be permissible to take into consideration a universe subjected according to laws not yet scientifically recognized (the big public is not aware) to "*a phenomenon of mutational frictions*" thanks to which the two systems (invisible and visible) interact on a condition: *The sustenance of frictions = the perpetual motion at every point of the visible whole with respect to both itself and the invisible.*

→ The creation and/or loss of a given portion of energy of the visible and/or observable universe should not be a kind of taboo.
On the Contrary, it is an existing reality caused constantly and simultaneously by the same phenomenon as illustrated by this new discipline (with enough empirical material evidences) called:

The universal mutation

A discipline that deals with the notion of motion and frictions inseparable from the visible whole, calling into question the supposed impossibility of perpetual motion; a reality of which neither relativity nor quantum mechanics gives much precision, which nevertheless exists and unites them.

HYPOTHESIS

2. The phenomenon of mutational frictions

(Depicting how frictions subsist along with the motion of visible bodies instead of annihilating).

2.1. The paradoxes of a visible body in motion.

* (The paradox of a circle that rolls on another stationary circle)

Let be a dynamic system with frictions that both destroy and allow simultaneously movement (slippage decrease – rolling increase), of two different circles A and B with respective radii R_A and $R_B = k\, R_A$; circle A rolls (inside or outside) on a stationary circle B.

Note: ''a body comprises matter generally represented as a sphere (a three-dimensional figure analogue to a two-dimensional circle and to a one-dimensional point) either at the infinitely large scale or at the infinitely small scale''.

With reference to **figure 1**

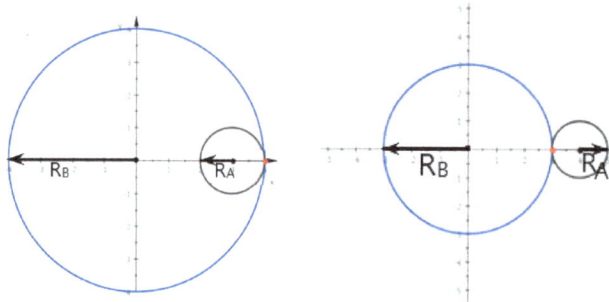

Wheel slip α = (travelled distance with slip) / (total travelled distance)

$k = (2\pi R_B)/(2\pi R_A)$ for $R_B = k R_A$

The circle A achieves:

- ''$(1-\alpha)k$'' number of rotations (turns) to travel the circumference ($2\pi R_B$) of circle B.
- And ''1'' revolution around the center of circle B.

Paradoxically to the usually erroneous answer generally granting ''$(1-\alpha)k$'' number of rotations to circle A during its journey around the stationary circle B, with any speed of the moving circle A:

- For a circle A moving inside the stationary circle B, circle A always achieves ''$(1-\alpha)k$'' number of rotations ''-1 turn'' to regain its starting position due to its revolution which is in the opposite direction as that of its rotation (it logically means that the travelled **distance shrinks** and the elapsed **time speeds up** for a circle A).

- For a circle Λ moving outside the stationary circle B, circle A always achieves ''$(1-\alpha)k$'' number of rotations ''+1 turn'' to regain its starting position due to its revolution which is in the same direction as that of its rotation (it logically means that the travelled **distance enlarges** and the elapsed time **slows down** for a Circle A).

In generally speaking: ''any visible body moving in any orbit (at any scale, with no need to necessarily reach higher velocity like the speed of light, space and time respectively contract and dilate vice versa) appeals to the notion of imaginary time; it is all about orbital geometry''.

**   (The ARISTOTLE's wheel paradox)

Let be a dynamic system with frictions that both destroy and allow simultaneously movement (slippage decrease = rolling increase), of two parallel circles A and B (belonging to the same sphere C) with respective radii R_A and $R_B = k\, R_A$.

The sphere C rolls on any path of a given m length.

Note: ''a body comprises matter generally represented as a sphere (a three-dimensional figure analogue to a two-dimensional circle and to a one-dimensional point) either at the infinitely large scale or at the infinitely small scale''.

With reference to figure 2

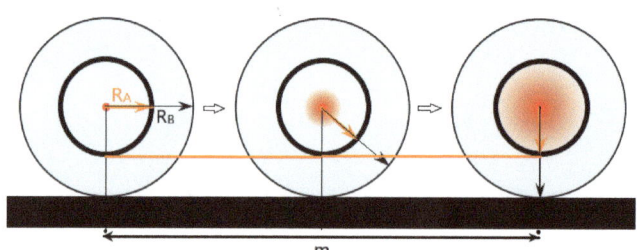

Wheel slip α = (travelled distance with slip) / (total travelled distance)

$$k = (2\pi R_B)/(2\pi R_A) \quad \text{for} \quad R_B = k\, R_A$$

Paradoxically, if k ≠ 1, one may think of [(1-α)m/ (2πR_A)] the achieved number of rotations by the circle A as being generally different from [(1-α)m/ (2πkR_A)] the achieved number of rotations by the circle B.

Yet, with any speed of the moving sphere C:

[(1-α) m/ (2πR_A)] = [(1-α) m/(2πkR_A)]   either for k = 1 or k ≠ 1.

- For parallel circles A and B of a single sphere C rolling on any path, the achieved numbers of rotations respectively by circle A and by circle B are always equal (it logically means that both **unequal circles A and B** have travelled the **same exact distance** at the **same exact rotational speed** during the **same elapsed period of time**, which turns out to be **illogic**).

In generally speaking: ''any moving visible body (at any scale, with no need to necessarily reach higher velocity like the speed of light, mathematical-conceptions of space and time are broken) appeals to the notion of singularity; it is all about motion''.

OBSERVATION

2.2. Analysis of simple machines

Reminder

-A circular wheel comprises an infinite number of radii (finite number of spokes of same length) that represent each the length of a lever arm (historically invented by the first world's civilization of Egyptians in Africa and somehow known generally as ARCHIMEDES lever); and the center of that wheel represents the fulcrum.

-A fixed-pulley changes the direction of forces without affecting their magnitudes (no matter the direction of a given force applied via a rope or a chain and/or a belt, the force's line of action is always continuously tangential to the wheel's circumference: ''$\beta = 90°$ is an angle between the force's line of action and the wheel's radius'').

-A net force F_{Fr} from passive and/or stress forces (due to slippage, elasticity, weight of the wheels, ambient temperature, vibrations and so on...) of the dynamic system has to theoretically be considered as negligible for the purpose of a simplified math analysis (systems below consist of ideal circular gears or wheels and/or gear wheels that are connected by means of ideal ropes or belts and/or chains).

-As long as a given force is continuously applied and transmitted tangentially on a given wheel's circumference, ''the wheel's radius just becomes a simple lever'' enabling the use of the notion of force to be more fundamentally preferable (instead of torque).

And the magnitude of that force (justifying interaction) is one among fundamental physical quantities that has to be considered along with the radius (justifying space) and the rotational speed (justifying the time-flow); to finally have a fundamentally simplified generic-notion of levers (useful in the rest of this theory).

-A use of indirect contact transmission (by means of chains connecting gear wheels, instead of direct contact by means of gears) is just to facilitate the learning and understanding of natural phenomena such as gravitation (mentioned in the rest of this theory).

With reference to **figure 3**

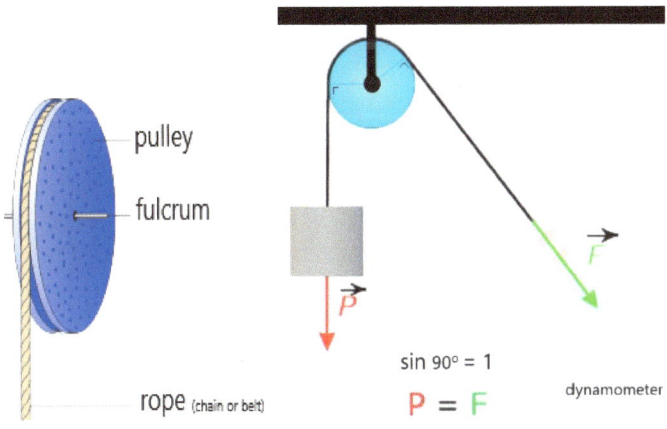

By definition for any ideal pulley, the wheel slip α=(travelled distance with slip)/(total travelled distance)≈0

Analysis

The present analysis depicts the notion of levers and obeys each of already recognized scientific principles (particularly the principle of momentum and/or energy conservation); so as to have a more useful and clear understanding of the variation of the following fundamental physical quantities:

''forces (resistive or motive), rotational speeds and lengths of the lever arm (radius, or circumference and/or number of teeth of a gear wheel)''.

Let (F, R, V) be respectively the magnitude of a force (motive or resistive) applied tangentially on a wheel's circumference, the wheel's radius, the wheel's rotational speed.

And (A, B) are two connected wheels that form a simple machine and have respective radii (R_A, R_B); they rotate at respective rotational speeds (V_A, V_B) due to the action of respective forces (F_A, F_B).

During the transmission of movement from wheel **A** to wheel **B** (vice versa) by means of indirect contact transmission, the variation of these three physical quantities (**force, rotational speed and radius**) is generally described as follows:

1st Axiom

During the transmission of movement from wheel **A** to wheel **B** (vice versa), if the magnitude of a force does not vary (same force at the input and output), the radius is always proportional to the rotational speed.

If $F_A = F_B$ = nonzero constant and F_{Fr} = zero constant; we have:
$R_A = k\,R_B$ and $V_A = V_B / k$ or $R_B = R_A / k$ and $V_B = k\,V_A$

→ $R_A\,V_A = R_B\,V_B$

With reference to **figure 4**

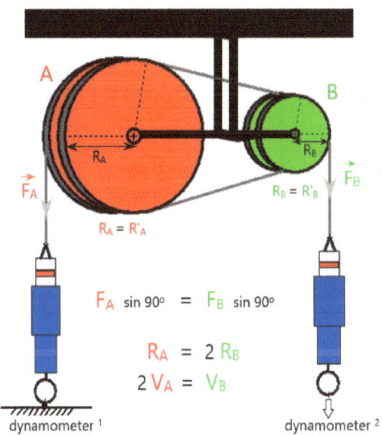

"A and B are wheels having radii R_A and R_B whose ratio **k** equals **2** (A and B are compound wheels connected via intermediary radii R'_A and R'_B whose ratio equals **2** as well); all wheels are capable of making a rotational movement around two respective centers separated by a given (not necessarily constant) distance; $\beta = 90°$ is always an angle between the force's line of action and the wheel's radius (no matter the direction of dynamometers and the disposition of the system as a whole); and a net force F_{Fr} from stress forces of the system is theoretically negligible".

2nd Axiom

During the transmission of movement from wheel **A** to wheel **B** (vice versa), if the rotational speed does not vary (same speed at the input and output), the force's magnitude is always proportional to the radius.

If $V_A = V_B$ = nonzero constant and F_{Fr} = zero constant; we have:
$F_A = F_B / k$ and $R_A = k\, R_B$ or $F_B = k\, F_A$ and $R_B = R_A / k$

→ $F_A\, R_A = F_B\, R_B$

With reference to **figure 5**

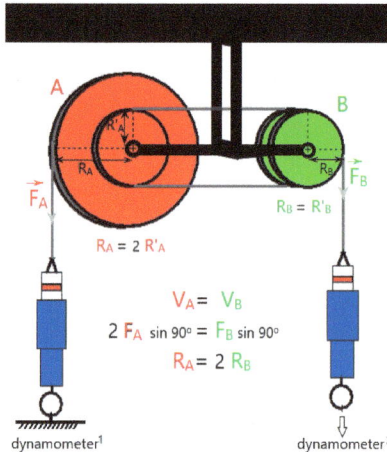

'' **A** and **B** are wheels having radii R_A and R_B whose ratio **k** equals **2** (**A** and **B** are compound wheels connected via intermediary radii R'_A and R'_B whose ratio equals **1**); all wheels are capable of making a rotational movement around two respective centers separated by a given (not necessarily constant) distance; $\beta = 90°$ is always an angle between the force's line of action and the wheel's radius (no matter the direction of dynamometers and the disposition of the system as a whole); and a net force F_{Fr} from stress forces of the system is theoretically negligible''.

3rd Axiom

During the transmission of movement from wheel **A** to wheel **B** (vice versa), if the **radius does not** vary (same radius at the input and output), **the force's magnitude is always proportional** to the rotational speed.

If $R_A = R_B$ = nonzero constant and F_{Fr} = zero constant; we have:
$F_A = F_B / k$ and $V_A = k\, V_B$ or **$F_B = k\, F_A$ and $V_B = V_A / k$**

→ $F_A V_A = F_B V_B$

With reference to **figure 6**

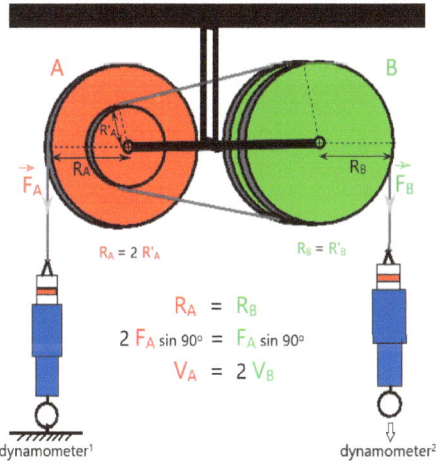

''**A** and **B** are wheels having radii R_A and R_B whose ratio **k** equals **1** (**A** and **B** are compound wheels connected via intermediary radii R'_B and R'_A whose ratio equals **2**); all wheels are capable of making a rotational movement around two respective centers separated by a given (not necessarily constant) distance; $\beta = 90°$ is always an angle between the force's line of action and the wheel's radius (no matter the direction of dynamometers and the disposition of the system as a whole); and a net force F_{Fr} from stress forces of the system is theoretically negligible''.

Notice

- When two among the three physical (nonzero) quantities (**force, rotational speed and radius**) do not vary, the other remaining quantity does not vary either.

- When each of the three physical (nonzero) quantities (**force, rotational speed and radius**) varies, one among the three physical (nonzero) quantities must vary **k** times (**k** being the product of **k'** and **k"**, with **k'** and **k"** being the respective numbers of times that the two other remaining physical quantities **inversely** vary).

With reference to **figure 7**

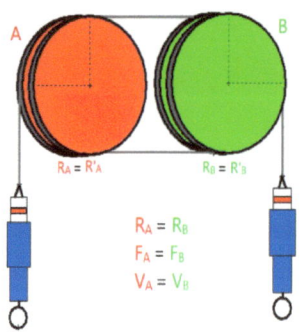

$R_A = R'_A$ $R_B = R'_B$
$R_A = R_B$
$F_A = F_B$
$V_A = V_B$

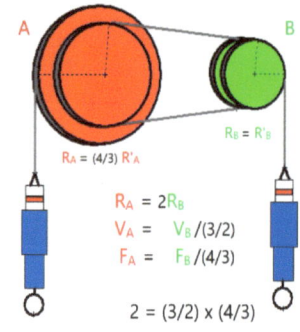

$R_A = (4/3) R'_A$ $R_B = R'_B$
$R_A = 2R_B$
$V_A = V_B /(3/2)$
$F_A = F_B /(4/3)$
$2 = (3/2) \times (4/3)$

- When taken into consideration, the notion of direct contact transmission via gears (affecting the direction of rotational movement, the elasticity, slippage, and so on...) affects the efficiency of the system in question, and the above cited axioms and notices are still obeyed.

- Whenever, in presence of any of the ideal systems mentioned above, there must be a conservation of momentum and/or energy; the input power equals the output power $F_A V_A R_A \sin 90° = F_B V_B R_B \sin 90°$.

And the above cited axioms and notices are all still obeyed when one (or two and/or three) among the three physical quantities is (or are) a zero constant.

For instance, during the transmission of motion, if the rotational speed becomes a zero constant, the system is at the equilibrium (at rest):

''If $V_A = V_B$ = zero constant and F_{Fr} = zero constant, no matter how radii and forces vary, $F_A V_A R_A \sin 90° = F_B V_B R_B \sin 90° = 0$''.

''And in case, any among these two physical quantities (the radius, the applied forces) equals zero, there is even no need to talk about transmission of movement''.

N.B:

Systems consisting of more than two wheels (*including all already known planetary gear-boxes provided with* a carrier that plays a role of a gear or a wheel and/or a lever, *and conforming to formula of* WILLIS: *"like a simple planetary, a planetary of Ravignaux, a planetary of Simpson and others ..."*) can all be simplified and represented (from the input to the output) as a system of two wheels obeying the above cited axioms and notices, no matter how complex the system in question is.

With reference to **figure 8**

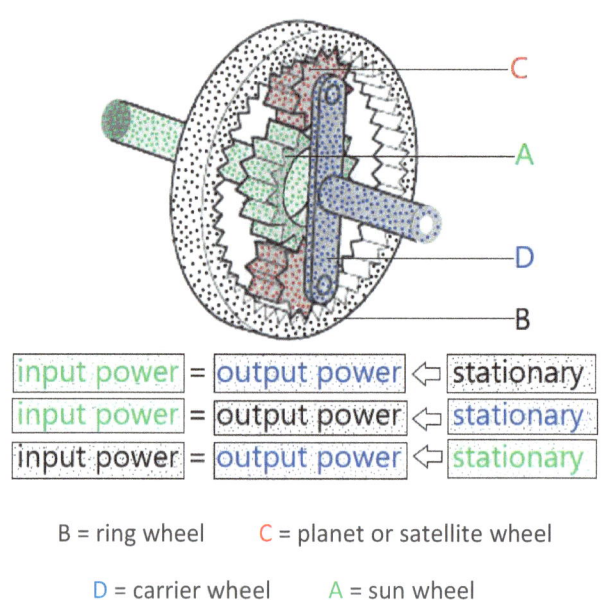

B = ring wheel   C = planet or satellite wheel

D = carrier wheel   A = sun wheel

WARNING!

The use of language literally easy and does not involve some complex mathematical notions (like the notions of infinitesimal calculus, **differential** equations, logarithms, vectors, metric tensors and others… needed to scientifically generalize complex motion-friction phenomenon of complex visible bodies) is to initially evoke this theory and its testability at the fundamental scale-level; so as to minimize any immoral monopolization of the power of knowledge or know-how.

Tested in real world

README: whoever and wherever you are, whatever your point of view is, the HAKIZA-1 doesn't care: ''the HAKIZA-1 creates energy''. Run your own test for a deeper understanding and embrace the facts.

During the game,

curious are those who take risk;
anxious are those who become the risk itself;
victorious are those who understand it and act on time,

to stay in the game.

Pendant le jeu,

curieux sont ceux qui prennent le risque;
anxieux sont ceux qui deviennent le risque en-soi,
victorieux sont ceux qui le comprennent et agissent à temps,

pour rester dans le jeu.

The cosmos

On the ground always in motion
with the around in collusion,
I found my inspiration.
This is just an oral
about the new era;
and the word sounds wonderful
for the world to be hopeful.

Le cosmos

Sur une île en ruine nucléaire,
croisa ta fine née en avril,
l'inspira la nouvelle ère.
Comme nul ne trompe son propre cœur,
Nil ne cesse de couler vers le Caire;
et l'eau demeure en mode circulatoire
pour que le monde ait de l'espoir.

Long live Burundi, Africa and the rest of the world.

Get ready to experience a singularly unprecedented invention that will strengthen your innate while clarifying your actual acquired.
Especially do not try to deny its doctrine of truth for fear of losing what you already possess; weigh the contents and make the right decision:

«Here is a HAKIZA and its invisible fulcrum, let's preserve the world».

EXPERIMENT

2.3. Experiment of double orbital friction

Reminder

-A net force F_{Fr} from passive and/or stress forces of the dynamic system has to theoretically be considered as negligible for the purpose of a simplified math analysis (systems below consist of ideal circular gears or wheels and/or gear wheels that are connected by means of ideal ropes or belts and/or chains).

- A minimization of wheel slip requires the use of gears or gear wheels specifically between **the** stationary ring wheel and the planet wheel, to obtain regular cycloidal systems for the purpose of efficiency.

With reference to **figure 9**

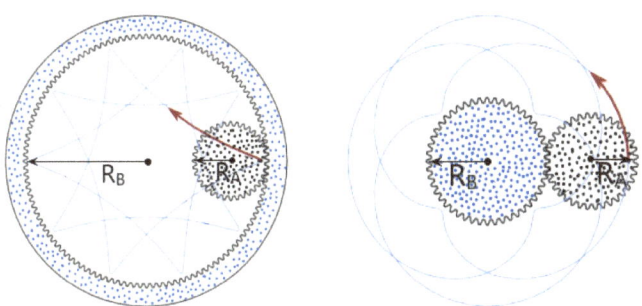

Wheel slip α = (travelled distance with slip) / (total travelled distance)

$$(1-\alpha)k = k \quad \text{for} \quad \alpha \approx 0$$

Analysis

This experiment consists in introducing wheels into a dynamic system that combines the two previously seen paradoxes (the paradox of a circle that rolls on another stationary circle & the Aristotle's wheel paradox); in order to constitute ''a PLANETARY TRANSMISSION WITH A COMMON COMPOUND PLANET WHEEL, A STATIONARY RING WHEEL, AN INPUT SUN WHEEL, AN OUTPUT SUN WHEEL AND A RESTORER WHEEL''.

- Let **A**, **B** and **C** be three wheels of respective radii R_A, R_B and R_C (double radius R_C and R'_C of a compound wheel **C**); and of respective rotational speeds V_A, V_B = zero constant and $V_C = V'_C$.

A and **B** have aligned centers on the same axle, so as to form a planetary system of three wheels (sun wheel **A** is connected by means of a belt or a chain to planet wheel **C** via radii R_A and R_C, the latter is a double compound wheel rolling with negligible slip "$\alpha \approx 0$" around a stationary ring wheel **B** via radii R_B and R'_C; in order to establish an orbital friction); just appealing to the combination of the two paradoxes mentioned above.

- Let then establish a double orbital friction by introducing another sun wheel A_1 within the previous dynamically orbital system, in order to constitute a system of four wheels A_1, A, B and C; of respective radii R_{A1}, R_A, R_B and R_C (triple radius R_C, R'_C and R''_C of a compound wheel C); and of respective rotational speeds V_{A1}, V_A, V_B=zero constant and $V_C = V'_C = V''_C$.

- *The ratio* $(R_C/R_A)/(R''_C/R_{A1})$ *must be different from* 1.
  Otherwise the below systems of multiple wheel would just be simplified into a simple machine of two wheels.
  And the three physical (nonzero) quantities (force, rotational speed and radius) would not vary during the transmission of motion.

A_1, A and B have aligned centers on the same axle; and B is in the middle of A and A_1.

So as to form a planetary system of four wheels (sun wheel A is connected by means of a belt or a chain to planet wheel C via radii R_A and R_C, the latter is a triple compound wheel rolling with negligible slip "$\alpha \approx 0$" around a stationary ring wheel B via radii R_B and R'_C, and finally connected by means of a belt or a chain to sun wheel A_1 via radii R_{A1} and R''_C; in order to establish a double orbital friction).

- Let finally add another wheel A_2 (a restorer-wheel belonging theoretically to the planetary system but placed physically outside of the same planetary system, and having a rotational speed V_{A2} and a radius R_{A2}) connected to the sun wheel A via R_{A2} and R'_A (so as to establish the same rotational speed and radius at the two ends of the dynamically orbital system, $V_{A2} = V_{A1}$ and $R_{A2} = R_{A1}$, then deduce the variation of the force's magnitude using dynamometers).

A_1, A, B, C and A_2 form a planetary system illustrated as follows:

With reference to **figure 10**

1st theorem

For any dynamically hypocycloidal system
(planet wheel **C** connected to sun wheel **A** via radii R_A and R_C, planet wheel **C** rolling with negligible wheel slip "$\alpha \approx 0$" around a stationary ring wheel **B** via radii R_B and R'_C), there is always a coefficient denoted **H⁻** determined as follows:

$H^-_A = (R_C / R_A) - (1-\alpha).(R'_C / R_B)$ that we name the hypocycloidal coefficient with respect to the sun wheel **A**.

The rotational speeds of sun wheel **A** and planet wheel **C** always vary with respect to each other according to the hypocycloidal coefficient.
$V_A = V_C . H^-_A$ for **V_B = zero constant**

With reference to **figure 11**

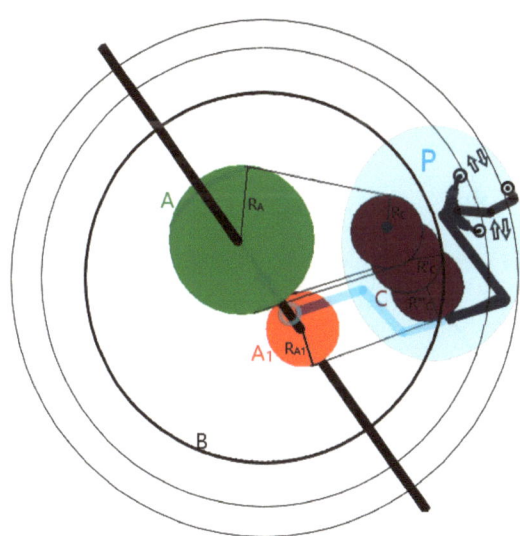

2nd theorem

For any dynamically epicycloidal system (planet wheel **C** connected to sun wheel **A** via radii R_A and R_C, planet wheel **C** rolling with negligible wheel slip "$\alpha \approx 0$" around a stationary ring wheel **B** via radii R_B and R'_C), there is always a coefficient denoted H^+ determined as follows:

$H^+_A = (R_C/R_A) + (1-\alpha).(R'_C/R_B)$ that we name the epicycloidal coefficient with respect to the sun wheel **A**.

The rotational speeds of sun wheel **A** and planet wheel **C** always vary with respect to each other according to the epicycloidal coefficient.

$V_A = V_C . H^+_A$ for V_B = zero constant

With reference to **figure 12**

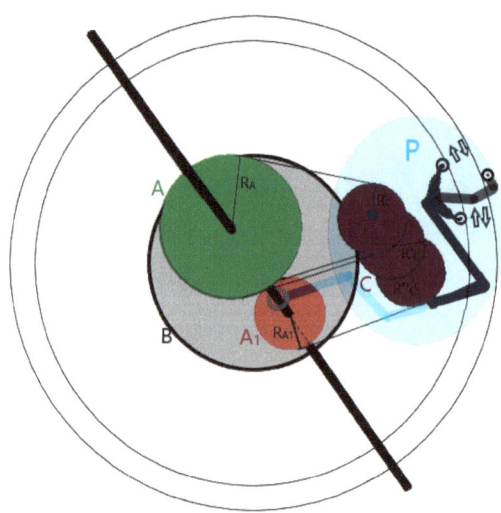

3rd theorem

Condition of double orbital friction:
$(R_A / R_C) / (R_{A1} / R''_C) \neq 1$

For any system respectively dynamically-hypocycloidal and dynamically-epicycloidal provided with double orbital friction (planet wheel C connected to A via radii R_A and R_C, planet wheel C rolling with negligible wheel slip "$\alpha \approx 0$" around a stationary ring wheel B via radii R_B and R'_C, and same planet wheel C connected to another sun wheel A_1 via radii R_{A1} and R''_C), there are always respectively two hypocycloidal coefficients denoted (H^-_A; H^-_{A1}) and two epicycloidal coefficients denoted (H^+_A; H^+_{A1}) determined as follows:

$H^-_A = (R_C/R_A) - (1-\alpha).(R'_C/R_B)$ and $H^+_A = (R_C/R_A) + (1-\alpha).(R'_C/R_B)$

$H^-_{A1} = (R''_C/R_{A1}) - (1-\alpha).(R'_C/R_B)$ $H^+_{A1} = (R''_C/R_{A1}) + (1-\alpha).(R'_C/R_B)$

The rotational speeds from wheel A to wheel C (vice versa) and from wheel C to wheel A_1 (vice versa) respectively vary as follows:

$V_A = V_C.H^-_A$ and $V_A = V_C.H^+_A$ for V_B = zero constant
$V_{A1} = V_C.H^-_{A1}$ $V_{A1} = V_C.H^+_{A1}$

The rotational speeds of sun wheels A and A_1 are always inversely proportional to their respective cycloidal coefficients.

$V_A/H^-_A = V_{A1}/H^-_{A1}$ (1)   and $V_A/H^+_A = V_{A1}/H^+_{A1}$ (2)

From (1) and (2); $V_A/H^\pm_A = V_{A1}/H^\pm_{A1}$ for V_B = zero constant.

Notice

- When we use gears instead of wheels, the direction of rotational motion is affected, and the above theorems are still obeyed.

Consequently, the hypocycloidal system for wheels (connected via belts or chains) becomes an epicycloidal system for gears (the revolution of a planet gear is in the same direction as that of the rotation of a sun gear).

- When we use gears instead of wheels, the epicycloidal system for wheels (connected via belts or chains) would become illusionary a hypocycloidal planetary system for gears (unfortunately, such system is generally not dynamic).

- When seeking more efficiency in real world, the use of either ''multiple planet wheel'' or ''a horizontal disposition of the system as a whole (in case of a planetary system having only one planet wheel, to compensate the weight of the carrier and its carried planet wheel)'' is recommended.

- When looking closely to both of the above planetary systems, there are two types of carrier or follower (in real world, it is not recommended to have both types of carrier in one system at once):

*A follower placed internally as a fixed-pulley rotating around the axle (less noisy and easy to build, recommended for the engineering of a powerful system either hypocycloidal or epicycloidal).

*A follower placed externally as a fixed-pulley moving in an orbit (too noisy and time-consuming to build, recommended for the purpose of seeking a theoretic profound understanding of a carrier's role for both types of carrier).

These present carriers just play a role of a dynamically orbital-fixed-pulley (**P**).

This specificity (only appearing in case of planetary systems with a common planet wheel) establishes a huge doubt, since a fixed-pulley does not play a role of a gear or a wheel and/or a lever; it just changes the direction of applied forces without affecting their magnitudes:

'' $F_{A1} V_{A1} R_{A1} \sin 90° = F_A V_A R_A \sin 90°$ is the input power still equal to the output power''?

With reference to **figure 13**

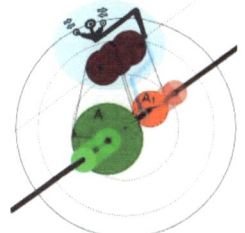

Epicycloidal ingoma

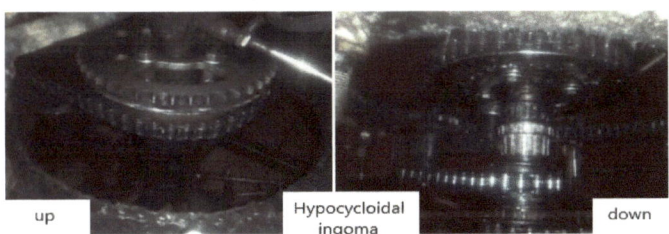

up   Hypocycloidal ingoma   down

RESULTS

2.3.1. HAKIZIMANA's paradox

(The duality of the radius within a ''**PLANETARY TRANSMISSION WITH A COMMON COMPOUND PLANET WHEEL, A STATIONARY RING WHEEL, AN INPUT SUN WHEEL, AN OUTPUT SUN WHEEL AND A RESTORER WHEEL**'')

(*)

For instance, a dynamically-orbital system consisting of four wheels (A_1 and A are sun wheels; C is a common compound triple planet wheel rolling with negligible slip "$\alpha \approx 0$" around a stationary ring wheel B) connected together so as to form a planetary transmission.

In case of hypocycloidal system, we suppose (optional) that $R_{A1}=R_C=2$ cm; $R_A = 4$ cm; $R_B = 8$ cm ; $R_C = R'_C = R''_C$; $R_{A1}=R_C=R_A/2 = R_B/4$ is the scale for diagram.

In case of epicycloidal system, we suppose (optional) that $R_{A1}=R_C=2$ cm; $R_A = 4$ cm; $R_B = 6$ cm; $R_C = R'_C = R''_C$; $R_{A1}=R_C=R_A/2 = R_B/3$ is the scale for diagram.

Let's verify the condition of double orbital friction

$(R_A / R_C) / (R_{A1} / R''_C) \neq 1$ ↔ $(4/2) / (2/2) = 2 \neq 1$

D.O.F. Condition is verified.

→ The cycloidal coefficients are determined according to the **1st and 2nd theorems**:

1st case: hypocycloidal system

$$H^-_A = (R_C/R_A) - (1-\alpha).(R'_C/R_B) \leftrightarrow H^-_A = 1/4$$

$$H^-_{A1} = (R''_C/R_{A1}) - (1-\alpha).(R'_C/R_B) \leftrightarrow H^-_{A1} = 3/4$$

For $V_A = V_C \cdot 1/4$ and $V_{A1} = V_C \cdot 3/4$

2nd case: epicycloidal system

$$H^+_A = (R_C/R_A) + (1-\alpha).(R'_C/R_B) \leftrightarrow H^+_A = 5/6$$

$$H^+_{A1} = (R''_C/R_{A1}) + (1-\alpha).(R'_C/R_B) \leftrightarrow H^+_{A1} = 4/3$$

For $V_A = V_C \cdot 5/6$ and $V_{A1} = V_C \cdot 4/3$

→ The rotational speeds of sun wheels A and A_1 are determined according to the **3rd theorem**:

1st case: hypocycloidal system

$$V_A/H^-_A = V_{A1}/H^-_{A1} \leftrightarrow V_A = V_{A1}/3$$

2nd case: epicycloidal system

$$V_A/H^+_A = V_{A1}/H^+_{A1} \leftrightarrow V_A = 5V_{A1}/8$$

(**)

THE HAKIZA METHOD

The procedure is about depicting the operational mode (respectively at rest and in motion) of the two cited above planetary systems (respectively hypocycloidal and epicycloidal) using the combination of axioms and theorems cited in the present theory.

At the equilibrium (at rest):

Both of the systems can be considered as a system of multiple wheels ''simplified into a system of two wheels'' that only require the axioms to be understood.

Note that the stationary wheel is theoretically meaningless, at rest.

Once the movement is initiated (in motion) :

At this stage, that system of two wheels requires actually the combination of both axioms and theorems in order to be understood.

Note that the stationary wheel is practically a game changer, once in motion.

At the equilibrium (at rest):

The sun wheel **A** is connected to the intermediate planet wheel **C** via R_A and R_C; the stationary ring wheel B can be ignored (cycloidal systems do not show up, at rest); and the intermediate planet wheel **C** is connected to another sun wheel A_1 via R_{A1} and R''_C.

For both hypocycloidal and epicycloidal systems, if we simplify the above cited system into a system of two sun wheels **A** and A_1, physical quantities vary as follows:

The radii of sun wheels **A** and A_1 vary with respect to each other according to $(R_A / R_C) \cdot (R''_C / R_{A1})$ coinciding with the ratio resulting from the condition of double orbital friction $(R_A / R_C) / (R_{A1} / R''_C)$.

$(R_A / R_C) \cdot (R''_C / R_{A1}) = (4/2) / (2/2) = 2$

$\rightarrow R_A = 2 R_{A1}$

Aaccording to axioms of this theory:
''If $V_A = V_B$ = zero constant and F_{Fr} = zero constant, no matter how radii and forces vary, $F_A V_A R_A \sin 90° = F_B V_B R_B \sin 90° = 0$''.

No surprise, it is the case, we have:
$V_A = V_{A1}$ = zero constant
F_{Fr} = zero constant
$F_A = F_{A1}$ and $R_A = 2R_{A1}$
for $F_{A1} V_{A1} R_{A1} \sin 90° = F_A V_A R_A \sin 90° = 0$.

With reference to **figure 14**

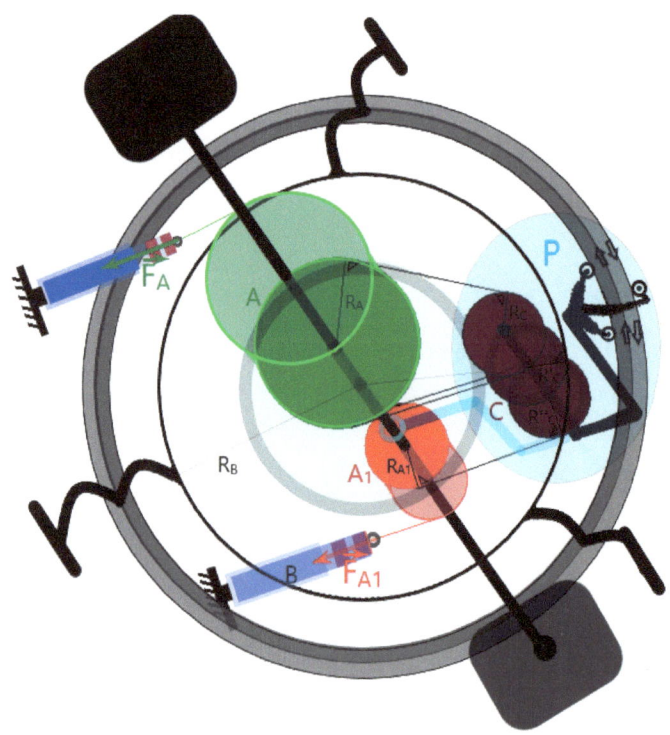

<u>AT THE EQUILIBRIUM</u> (AT REST): $F_A = F_{A1}$ $R_A = 2R_{A1}$ $V_A = V_{A1} = 0$

$$F_{A1} V_{A1} R_{A1} \sin 90° = F_A V_A R_A \sin 90° = 0$$

Once the movement is initiated (in motion):

The sun wheel **A** is connected to the intermediate planet wheel **C** via R_A and R_C; the stationary ring wheel **B** must be taken into consideration (once in motion; cycloidal systems show up, the ring wheel **B** is connected to the planet wheel **C** via R_B and R'_C); and the intermediate planet wheel **C** is connected to another sun wheel A_1 via R_{A1} and R''_C.

Let assume that the starting point of movement is the same for both systems (hypocycloidal and epicycloidal), by considering '' a small wheel A_1'' as a driving wheel and ''a large wheel **A**'' as a driven wheel.

For both (respectively) hypocycloidal and epicycloidal systems, if we simplify the above system into a system of two sun wheels **A** (output) and A_1 (input), physical quantities vary as follows:

The rotational speed of sun wheels **A** and A_1 must inversely vary proportionally to their cycloidal coefficients (according to theorems of this theory).

$V_A / H^-_A = V_{A1} / H^-_{A1} \leftrightarrow V_A = V_{A1}/3$ (hypocycloidal)

$V_A / H^+_A = V_{A1} / H^+_{A1} \leftrightarrow V_A = 5V_{A1}/8$ (epicycloidal)

Aaccording to the combination of theorems and axioms of this theory, once in motion, physical quantities vary as follows:

(Hypocycloidal system)

''If $F_A = F_{A1}$ = nonzero constant and F_{Fr} = zero constant; we have: $V_A = V_{A1}/3 \neq 0$ and we would consequently have $R_A = 3 R_{A1} \neq 0$ for $F_{A1} V_{A1} R_{A1} \sin 90° = F_A V_A R_A \sin 90°$ ''.

But it surprisingly not the case, we still have $R_A = 2R_{A1}$ and consequently $F_{A1} V_{A1} R_{A1} \sin 90° \neq F_A V_A R_A \sin 90°$.

(Epicycloidal system)

''If $F_A = F_{A1}$ = nonzero constant and F_{Fr} = zero constant; we have: $V_A = 5V_{A1}/8 \neq 0$ and we would consequently have $R_A = 8 R_{A1}/5 \neq 0$ for $F_{A1} V_{A1} R_{A1} \sin 90° = F_A V_A R_A \sin 90°$ ''.

But it surprisingly not the case, we still have $R_A = 2R_{A1}$ and consequently $F_{A1} V_{A1} R_{A1} \sin 90° \neq F_A V_A R_A \sin 90°$.

With reference to **figure 15**

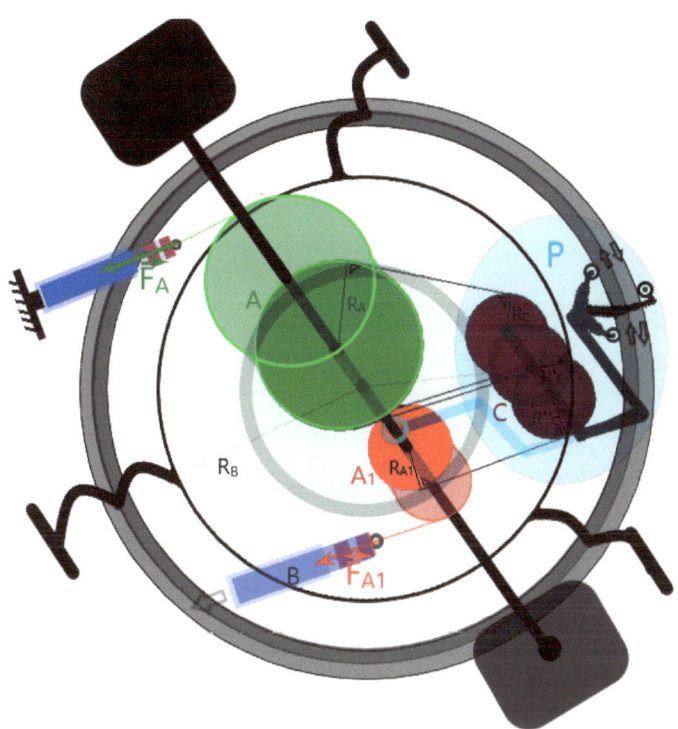

IN MOTION (hypocycloidal system):

$F_A = F_{A1}$ $R_A = 2R_{A1}$ $V_A = V_{A1}/3$

$F_{A1} V_{A1} R_{A1} \sin 90° \neq F_A V_A R_A \sin 90°$

IN MOTION (epicycloidal system):

$F_A = F_{A1}$ $R_A = 2R_{A1}$ $V_A = 5V_{A1}/8$

$F_{A1} V_{A1} R_{A1} \sin 90° \neq F_A V_A R_A \sin 90°$

-At the equilibrium (at rest), for both hypocycloidal and epicycloidal systems, the results would obey the principle of conservation of momentum and/or energy.

-In motion, for both hypocycloidal and epicycloidal systems, we obtain astonishingly two superposed radii (invisible or virtual radius, and another visible or real radius) simultaneously appearing at once and allowing the cited planetary systems to challenge the principle of energy conservation (once the two superposed radii are combined):

If sun wheel A_1 is considered as the driving wheel, we have:
$R_{real} = R_A = 2 R_{A1}$ and $R_{virtual} = R_A = 3 R_{A1}$ for $V_A = V_{A1}/3$ (hypocycloidal)
$R_{real} = R_A = 2 R_{A1}$ and $R_{virtual} = R_A = 8 R_{A1}/5$ for $V_A = 5V_{A1}/8$ (epicycloidal)

If sun wheel A is considered as the driving wheel, we have:
$R_{real} = R_{A1} = R_A/2$ and $R_{virtual} = R_{A1} = R_A/3$ for $V_{A1} = 3 V_A$ (hypocycloidal)
$R_{real} = R_{A1} = R_A/2$ and $R_{virtual} = R_{A1} = 5 R_A/8$ for $V_{A1} = 8V_A/5$ (epicycloidal)

Any of the dynamically orbital systems is at once both centrifugal in case ($R_{virtual} > R_{real}$) and centripetal in case ($R_{virtual} < R_{real}$).

Paradoxically, once in motion, the dynamically orbital system decides to act differently by creating simultaneously a new virtual radius (mathematically measurable but physically non-descriptible) while keeping its real radius (mathematically measurable and physically descriptible).

INTERPRETATION

2.3.2. The INGOMA

N.B: *etymologically taken from the national language of* **BURUNDI** "KIRUNDI", INGOMA *means: "drum, a symbol of power and reign"; it is a countable name both in singular and plural, a superposition of states deserving the designation of a planetary transmission of* **HAKIZIMANA**.
"UWUHARIRA KO KADYENDA ARI IGOMA YIVUZA Y'UBURUNDI, ARAZA DUHAYE AZODUSANGA".

During the transmission of movement via an ingoma, the interpretation of the variation of physical quantities depends on the HAKIZIMANA's paradox (the double radius within an ingoma: ''a virtual radius $R_{virtual}$ and real radius R_{real} observed either at the input or output''), which makes the transmission of movement completely different from what we observe in any system of machines including all already known traditional planetary gear-boxes.

Hence, we necessarily owe ourselves the gratitude to make the whole world discover what it really is:
''By making visible the virtual radius mathematically measurable $R_{virtual}$; and consequently re-establish same physical (nonzero) quantities (''at the input'' as well as ''at the output'') of two among the three fundamental physical quantities (force, rotational speed and radius), to finally deduce the variation of the remained physical quantity''.

Accordingly to "(*) and (**)", we provide any rotational speed ($V_{A1}>0$, at the input) to the sun wheel A_1 (having a radius R_{A1}).

And at the output, we add another wheel A_2 (a restorer-wheel placed outside of the planetary system, connected to the sun wheel at the input or output, having a rotational speed V_{A2} and a radius R_{A2}) connected to the sun wheel A (having two different radii: "a virtual radius $R_{virtual} = R'_A$ and a real radius $R_{real} = R_A$") via R_{A2} and $R_{virtual}$.

So as to establish the same rotational speed and radius ("at the input" and as well as "at the output"; $V_{A2}=V_{A1}$ and $R_{A2}=R_{A1}$); and finally deduce the variation of the force's magnitude.

Note that even if the restorer wheel is placed physically outside of the planetary system, as long as its speed and its radius are being determined in regard to the virtual radius, it theoretically belongs to that same planetary system (at given extent).

With reference to **figure 16**

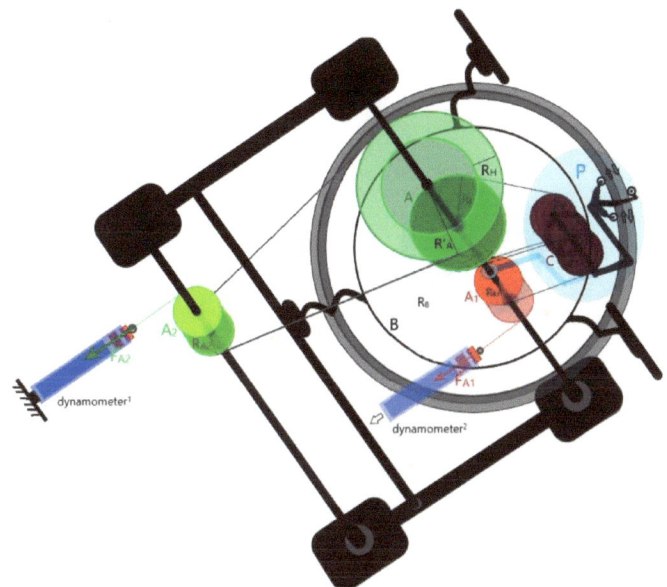

$R_{real} = R_A = 2R_{A1}$

$R_{virtual} = R'_A = 3 R_{A1}$

$R_H = |R_{real} - R_{virtual}|$

$n = R_{real}/R_{virtual} = 2/3$

CENTRIFUGAL HYPOCYCLOIDAL INGOMA

$n\, F_{A1} = F_{A2}$ for $n \in\,]0, 1[$ and $1/n \in\,]1, +\infty[$

$V_{A1} = V_{A2}$

$R_{A1} = R_{A2}$

$F_{A1}\, V_{A1}\, R_{A1} \sin 90° > F_{A2}\, V_{A2}\, R_{A2} \sin 90°$

With reference to **figure 17**

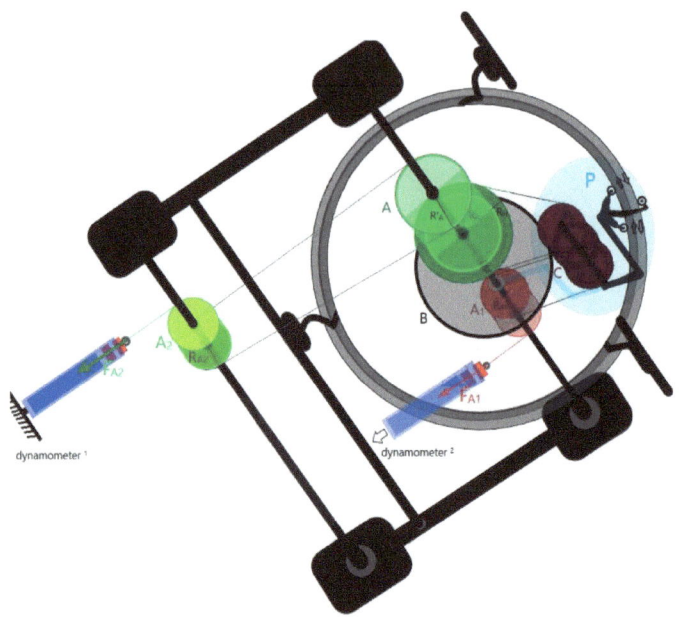

$R_{virtual} = R'_A = 8R_{A1}/5$

$R_{real} = R_A = 2R_{A1}$

$R_H = |R_{real} - R_{virtual}|$

$n = R_{real}/R_{virtual} = 5/4$

CENTRIPETAL EPICYCLOIDAL INGOMA

$n\, F_{A1} = F_{A2}$ for $n \in\,]1, 2[$ and $1/n \in\,]½, 1[$

$V_{A1} = V_{A2}$

$R_{A1} = R_{A2}$

$F_{A1}\, V_{A1}\, R_{A1} \sin 90° < F_{A2}\, V_{A2}\, R_{A2} \sin 90°$

We deduce that an ingoma is a "**planetary transmission with a common compound planet wheel**" comprising in principle five wheels; all wheels being arranged and connected to each other by means of belts or chains and/or connected directly in case of gears; so as:

- To comply with the condition of double orbital friction $(R_A/R_C)/(R_{A1}/R''_C)$ must be different from 1. R_C, R'_C, R''_C are radii of a necessarily triple common compound-wheel called planet or satellite (**C** connected to **A** via R_A and R_C, **C** rolling with negligible slip on a stationary wheel **B** via R_B and R'_C, and **C** connected to A_1 via R_{A1} and R''_C).

R_A is the radius of the sun wheel **A**, and with the stationary wheel **B**, they have aligned centers on one side of the axle.

R_{A1} is the radius of another sun wheel, and with the stationary wheel **B**, they have aligned centers on another side of the same axle.

- To allow movement within the system in question.

- To obey all laws already recognized scientifically (conservation of energy and/or momentum), once at the equilibrium (at rest).

- To obey theorems mentioned above, once it is in motion, resulting in a double nature that manifests itself by a paradoxically unexpected appearance of two superposed radii $R_{virtual}$ and R_{real} (in a simultaneously superposed action).

We provisionally call the ratio of duality ''$n = R_{real} / R_{virtual}$ and its inverse $1/n = R_{virtual} / R_{real}$'', (according to the starting point of the movement).

R_{real} is the real radius appearing both at the equilibrium (at rest) and once the movement is initiated (in motion).

And $R_{virtual}$ is the invisible radius mathematically measurable only after the initiation of movement.

If $n = 1/n = 1$, we are not in presence of an ingoma, and the principle of conservation of momentum and/or energy is obeyed.

And the absolute value $R_H = |Rreal - Rvirtual|$ is astonishingly a completely both invisible and visible radius that we name: ''THE HAKIZA-RADIUS''.

A. Mutated or mutational physical quantity

According to **figure 16** and **figure 17**.

-We generally notice that the force's magnitude varies (appealing to the ratio of duality ''n'') as follows:

The force's magnitude (at the input) is either (depending on the starting point of the movement) divided or multiplied by ''n'' factor (at the output), while the rotational speed and radius remain the same, during the transmission of movement.

-Consequently, increasing or decreasing the radius of one of the wheels (at the input or output, so as to re-establish the same force's magnitude and same rotational speed), it allows us to generally deduce the variation of radius (appealing to the ratio of duality ''n''):

The radius (at the input) is either (depending on the starting point of the movement) divided or multiplied by ''n'' factor (at the output), while the force's magnitude and rotational speed remain the same, during the transmission of movement.

- Subsequently increasing or decreasing the radius of one of the wheels (at the input or output, so as to re-establish the same force's magnitude and same radius), it allows us to generally deduce the variation of rotational speed (appealing to the ratio of duality ''n''):

The rotational speed (at the input) is either (depending on the starting point of the movement) divided or multiplied by ''n'' factor (at the output), while the force's magnitude and radius remain the same, during the transmission of movement.

In other words, more ''one among the two sun wheels and the stationary wheel become significantly large'' compared to other wheels, more the ratio of duality increases for an epicycloidal system, more the ratio of duality inversely increases for a hypocycloidal system.

It is obviously tempting to constitute a kind of hybrid system with both epicycloidal (using gears instead of gear wheels to transmit movement by means of a direct contact from the common planet wheel to the significantly large sun wheel, at the output side of the planetary system; it affects the direction and speed) and hypocycloidal (using gear wheels to transmit movement by means of chains from the small sun wheel to the common planet wheel, at the input side of the same planetary system); unfortunately we end up either with a system having low efficiency or a system generally no-dynamic.

We are right-handed … with naked eyes and hand holding the input shaft and the output shaft having same radius... it is so crazy to feel and watch our own left hand pulling our own right hand at the same rotational speed rate.

Are we in presence of an artifical so-said "dark matter, vacuum or dark energy, and black-white twin hole (wormhole)"?

Yet one thing is certain, we are in presence of a missing and/or excess of a mathematically measurable physical quantity (the so-called dark gravity).

METAPHYSICS: Unlike most scientists, We don't even need to directly tackle the beast (dark matter and/or dark energy); since by a simple thought-experiment based on the notion of the referential-frame: "the physically descriptible invisible" is no longer invisible, it becomes "the visible" which further reaffirms via mathematical recurrence of **POINTCARE** the existence of "the invisible" to which one must refer in order to determine and describe the meaning and the very cosmic essence of the word "visible".

The scientist-ego-paradox: "it is bravely amazing and recommended for all of well-thinking beings, to directly tackle dark gravity and excite ourselves as much as we want either with chromo-dynamics or with cosmology (exacerbating the enigma of dark gravity within inefficiently-artificial collider-accelerators and telescopes); *this is an attempt to satisfy our own insatiable ego, to finally prove and reaffirm our own misunderstanding of dark gravity*".

However, there is a way to remedy the above paradox according to **DESCARTE:** *"I think therefore I am"*.

There are no objections of admitting in the very valid sense of the term "thought (as a portion of mathematically measurable and physically non-descriptible information within imaginary **HAWKING** time, capable of becoming descriptible once emerged in real time, vice versa)" that as long as it is still invisible in nature, it can gradually exceed the speed of light to infinity, according to the limits imposed by the one who is "thinking (as process)".

The scientist-mindset-remedy: "information moves at a given limited speed in real time, to move at a given unlimited speed within the imaginary time, and with no need to be either a neuron-scientist or a cosmologist (experiencing the enigma of dark gravity within our own brain considered as both an efficiently-natural collider-accelerator and telescope); *this is an attempt to satisfy our own insatiable mind, to finally prove and reaffirm our own resemblance to dark gravity*".

For instance in real time, any information (moving at the speed of light) from the so-called big-bang must travel several billions of light-years to reach us. Yet, within imaginary time, the information in a human-being's brain can even go in a split second beyond that big-bang and speculate on what happened... moreover, even if the calculations still need to be corrected, the universe itself is somehow expanding faster than the speed of light (accurately, it is mathematically the invisible space metric expanding "faster than light").

In short, the reformulation according to **HAKIZIMANA:**
"the invisible exists therefore I am the fruit of it".
And as everyone would say: "You recognize the tree by its fruit".

We therefore must call "the invisible" by its own and only valuable name "**the invisible** = any portion of information within imaginary time (moving faster than the speed of light within and beyond the visible) and capable of becoming visible in real time according to mutational laws (described in the rest of the present theory)".

Physically the invisible is not a state (solid, or liquid, or gaseous and/or plasma) but rather mathematically say a limit stage of the visible whole (thus, another non-physical system that both simultaneously converges at any point of the physical system towards the micro and diverges at every point of the physical system towards the macro; since it is necessarily faster than the speed of light to finally establish the singularity ... *in other words,* the math-physics-unification = the information *).*

In any case, despite the interconnection between the fundamental physical quantities (force, rotational speed, radius) stipulated by all theories (scientifically recognized, in order to prove the existence of energy and/or momentum conservation within a given visible dynamic system) during a transmission of movement ; each of these cited three physical quantities (once in presence of a transmission of movement via an ingoma) evolves (accordingly to the HAKIZA RADIUS $|R_{real} - R_{virtual}|$) independently of the two others that remain.

During a transmission of movement, an ingoma is not concerned with a simple transfer of the initial values of quantities mentioned above; it is at once a matter:

- of creation ''n'' times of its initial values from what we do not see (the invisible).
- of loss ''n'' times of its initial values towards what we do not see (the invisible).

The invisible

When it comes to the notion of ''the invisible'' synonym of ''singularity'', the notions of referential frames, dimensions and degrees of freedom become meaningless and useless.

Consequently the transmission of energy via an ingoma is not anymore a simple transformation or transfer of physical given quantities.

It's convenient therefore to say that it is a mutation of physical quantities from the invisible to the visible (vice-versa) and we definitely call ''the ratio of duality $n = R_{real} / R_{virtual}$'' a BASIC MUTATIONAL FACTOR.

Mutational determinacy principle

The HAKIZA-RADIUS is both visible and invisible (any attempt to exceed the HAKIZA-RADIUS, results in the reappearance of the conservation of energy accordingly to the excess); with an astonishing principle (at any scale) that we name the principle of mutational determinacy and that we theoretically state as follows:

"Any maximum precision of the mutational variation of one among the three fundamental physical quantities (force, rotational speed and radius) results in the invariance of the other two remaining physical quantities".

Hence, every time the movement within an ingoma is initiated, we are in presence of physical quantities in a mutated form (a mutated force or rotational speed and/or radius).

Due to the double orbital friction occurring during some well-structured gravitational movements in the visible world (at any scale), we therefore deduce a phenomenon that gives a rise to the so-called dark gravity, finally allowing the mutation of a physical quantity from the invisible to the visible (vice-versa); and that we name:

''THE PHENOMENON OF MUTATIONAL FRICTIONS''.

Mutational transmission principle

Regardless of the referential frame and/or the dimension, thanks to this phenomenon illustrated above, there is a new principle of mutational transmission of motion from which nothing can escape, even the momentum and/or energy conservation principal collapses:

''**Within any dynamically orbital system, the visible whole is created and lost in the invisible; it is just a mutation of which only the trigger of the movement decides its beginning, its magnitude and its end**''.

Hence, every time the movement within an ingoma is initiated, we are in presence (as a result) of a mutational energy violating the principle of energy-conservation, and once involved in a given transformation into any other form of energy, that other form of energy also becomes mutational at a given extent (it does not give a damn to ''the preceding violation of energy-conservation principle'').

2.3.3 The HAKIZA

N.B: *etymologically taken from the national language of* BURUNDI "KIRUNDI", HAKIZA *is an abbreviation for* HAKIZIMANA, *a proper name that means:* "*only* GOD is *the savior*"; *deserving the designation of multiple* HAKIZIMANA's *planetary transmissions connected together in series (it is a set of multiple ingoma).*

B.1 Mutational factorization

Mutational factorization (1st mutational law)

> Any HAKIZA consisting of "p" number of ingoma (arithmetically linked in series) having each a basic mutational factor "n", the total mutated or mutational quantity of that HAKIZA always evolves according to the factor "N" (a product of all the basic mutational factors).
>
> The equation of a mutation:
>
> $X_m = X \cdot N$ and its symmetry $X_m = X \cdot (1/N)$
>
> with $N = (n_1 \cdot n_2 \cdot \ldots \cdot n_p)$ and $n = R_{real} / R_{virtual}$.

" $X_m = X \cdot N$ and/or $X_m = X \cdot (1/N)$ " is the equation of a mutation.

"n" is the duality ratio between the real radius and the virtual radius (basic mutational factor).

"p" is the number of ingoma connected (arithmetically) in series.

"N" is the total mutational factor within a HAKIZA.

"X" is a physical quantity with enough energy needed to trigger a movement.

And "X_m" is a mutated or mutational quantity that expresses the magnitude of a mutation.

E.g.:

For instance, **two** ingoma (with wheels whose radii are taken with reference to **figure 16 and figure 17** having respectively a basic mutational factor n = 2/3 and n = 5/4) connected together so as to form a system of two planetary transmissions in series (the driven wheel of the 1st ingoma is connected to the driving wheel of the 2nd ingoma).

"The total mutational factor is the product of the basic mutational factors of each ingoma of the system in question".

→ $N = n^2 = (2/3)^2$ for the two connected hypocycloidal ingoma of same type.

→ $N = n^2 = (5/4)^2$ for the two connected epicycloidal ingoma of same type.

→ And $N = n_1 \cdot n_2 = (2/3) \times (5/4)$ for the hypocycloidal system connected to the epicycloidal ingoma.

B.2 Mutational invariance

Mutational invariance (2nd mutational Law)

> Any ingoma whose radius of the stationary wheel (orbit) is considered as infinitely larger compared to the radii of the remaining rotary wheels (sun, planet and/or satellite), that ingoma always have a basic mutational factor ''n'' approximately equal to one.
>
> $$R_{real} \approx R_{virtual} \quad \leftrightarrow \quad X_m \approx X$$

E.g.:

For instance, **an** ingoma (with wheels whose radii are taken with reference to **figure 16** and **figure 17; exceptionally consisting of a stationary wheel B having a radius R_B considered as infinitely large compared to radii of rotating wheels**) must have:

"A basic mutational factor approximately equal to one ".

→ $N = R_{real}/R_{virtual} = 0.99999...989...$ for the hypocycloidal ingoma.

→ $N = R_{real}/R_{virtual} = 1.00000...010...$ for the epicycloidal ingoma.

B.3 Mutational singularity

The Mutational Singularity (3rd mutational Law)

Any single ingoma in any given HAKIZA has a basic mutational factor ''n'' that always increases or decreases infinitely within the singularity.

The basic mutational factor ''n'' is always defined in the range $]0, 1[\cup]1, 2[$.

And its inverse ''1/n'' is always defined in the range $]½, 1[\cup]1, +\infty[$.

-The value of ''n'' is always strictly above zero and strictly less than one, for a hypocycloidal system.

-The value of ''n'' is always strictly above one and strictly less than two, for an epicycloidal system.

E.g.: For instance, An ingoma (with wheels whose radii are taken with reference to **figure 16** and **figure 17**) must have :

"A basic mutational factor **n** defined in the range $]0, 1[\cup]1, 2[$; and its inverse **1/n** defined in the range $]½, 1[\cup]1, +\infty[$ ".

→ **n** = 2/3 for the hypocycloidal system and **n** = 5/4 for the epicycloidal system; and its inverse **1/n** = 3/2 for the hypocycloidal system and **1/n** = 4/5 for the epicycloidal system.

PROTOTYPE

C. The HAKIZIMANA's prototype machine

N.B.: THIS INVENTION IS UNDER PATENT PROTECTION. AND CAN ONLY BE EXPLOITED FOR PROFIT THROUGH A PARTNERSHIP OR LICENSE AGREEMENT WITH THE COMPANY HAKIZA-TECHNOLOGY (HATEC) BASED IN BURUNDI/AFRICA OR VIA THE SPONSORSHIP OF THE GOVERNMENT OF BURUNDI.

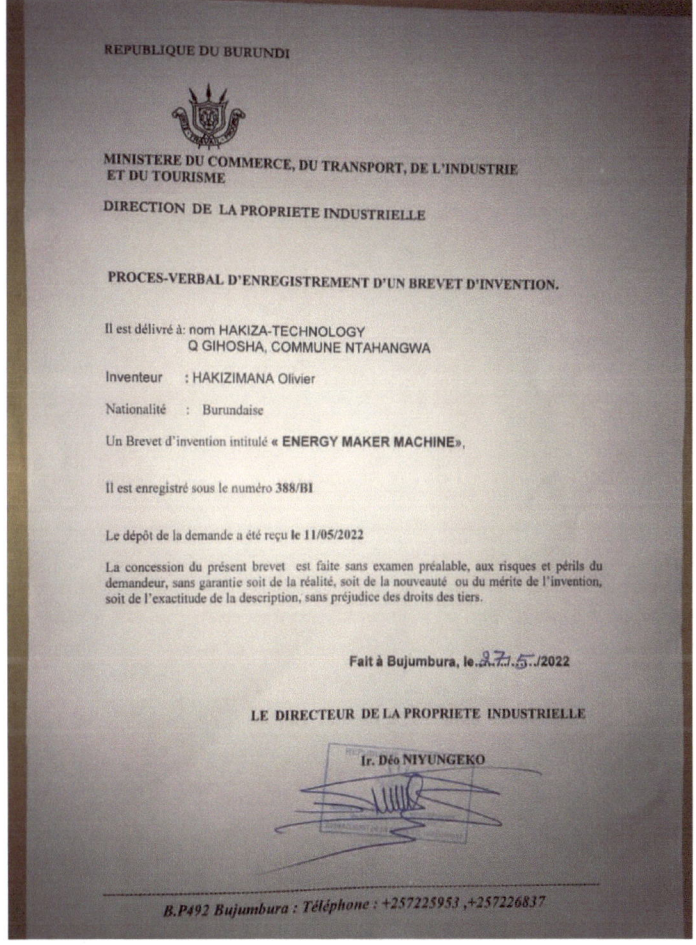

An international patent application has already been filed.

(ABSTRACT)

**ENERGY MAKER MACHINE
HAVING AN ELECTRIC MOTOR AND AN ALTERNATOR
JOINED VIA A PLANETARY TRANSMISSION HAVING A
COMMON COMPOUND PLANET WHEEL, A STATIONARY
RING WHEEL, AN INPUT SUN WHEEL, AN OUTPUT SUN
WHEEL, AND A RESTORER WHEEL.**

The present invention relates to a device that generates clean, inexhaustible and sustainable mechanical energy; which it then transforms into clean, inexhaustible and sustainable electrical energy.

The device has a system configuration mainly comprising a ''PLANETARY TRANSMISSION WITH A COMMON COMPOUND PLANET WHEEL'' placed in series between an electric motor and an alternator.
And unlike other devices comprising traditional planetary gear sets that simply transmit the torque accordingly to the rotational speed variation, the ''PLANETARY TRANSMISSION WITH A COMMON COMPOUND PLANET WHEEL'' increases by ''n'' factor the input torque and provides a higher output torque even if the rotational speed of the output shaft is restored and equal to the rotational speed of the input shaft.
This consequently allows this device to be uniquely climate-friendly by self generating its own mechanical energy, which it then transforms into clean, inexhaustible and sustainable electrical energy:

- First, by supplying electric energy from a DC electricity of a battery **(13)** to a control box **(15)**, the later is connected to an electric motor via its AC line **(16)** and its hall line **(17)**. The battery is useful only at startup and it can remain connected in order to recharge or be disconnected using a switch **(14)**.
- Second, by transforming the received electric energy into a mechanical energy; the electric motor **(18)** drives an input shaft **(12)** of the ''PLANETARY TRANSMISSION WITH A COMMON COMPOUND PLANET WHEEL **(19)**''.

- Third, by multiplying the input torque by ''n'' factor in order to obtain ''n'' times higher output torque while the input and output rotational speed remain the same; the ''PLANETARY TRANSMISSION WITH A COMMON COMPOUND PLANET WHEEL'' increases the received mechanical energy and transmits it via its output shaft **(10)** to an alternator.
- Fourth, by transforming the increased mechanical energy back into electric energy; the alternator **(20)** feeds energy back via its AC line **(21)** to an AC-DC regulator/rectifier **(22)**.
With that AC-DC regulator/rectifier connected via its DC line **(24)** to a battery in order to recharge it accordingly to the chosen configuration, to the electric motor control box in order to maintain the motor rotational speed even when the battery is disconnected, and as well as to an inverter **(23)**.
- Fifth, by harvesting the excess of usable and regulated electric energy via the inverter output **(25)**, since the produced electric energy from the alternator is much higher than what the electric motor needs to maintain its rotational speed and what the battery needs to recharge.

    According to the present invention, the device is industrially intended to self-produce clean, inexhaustible and sustainable energy; usable at zero emission in automotive devices, tractors, locomotives, aerospace devices, naval devices, robotic devices in general, as well as in power plants and electric generators.

(DESCRIPTION)

The present invention relates to a device that generates clean, inexhaustible and sustainable mechanical energy; which it then transforms into clean, inexhaustible and sustainable electrical energy.

Devices that need a very highly efficient torque transmission use a traditional planetary gear sets, like a simple planetary, or a planetary of SIMPSON having necessarily a ''common sun gear'', and/or a planetary of RAVIGNAUX having necessarily at once two ''sun gears'' and two connected ''planet gears''.
The traditional planetary transmission has in principle four interconnected wheels:
- A sun gear, randomly rotary or stationary depending on the desired configuration.
- One or multiple planet gears, necessarily rotary regardless of the configuration.
- A ring gear, randomly rotary or stationary depending on the desired configuration.
- A planet carrier, a gear randomly rotary or stationary depending on the desired configuration.
Traditional planetary transmission operates by effectively multiplying or dividing the initial torque by ''m'' factor while respectively decreasing or increasing the rotational speed by the same ''m'' factor; with that factor disappearing each time that the rotational speed is restored.

The present device comprises a ''PLANETARY TRANSMISSION WITH A COMMON COMPOUND PLANET WHEEL'' that increases by ''n'' factor the input torque and provides a higher output torque even if the rotational speed of the output shaft is restored and equal to the rotational speed of the input shaft.
And unlike other devices comprising traditional planetary gear set that simply transmits the torque accordingly to the rotational speed variation, with the initial motion being produced from mostly unclean and surely exhaustible energy resources; the present device uniquely self generates its own clean, inexhaustible and sustainable mechanical energy, which allows it to be climate-friendly.
The present device then transforms that mechanical energy into clean, inexhaustible and sustainable electrical energy thanks to a system configuration comprising mainly that ''PLANETARY TRANSMISSION WITH A COMMON COMPOUND PLANET WHEEL'' placed in series between an electric motor and an alternator.

The drawings of the present invention are briefly described as follows:
FIG.1a is a schematic representation of a HYPOCYCLOIDAL PLANETARY TRANSMISSION WITH A COMMON COMPOUND PLANET WHEEL, having an indirect contact by means of chains or belts between rotary wheels or gear-wheels, incorporating a family member of the present invention.
FIG.1c is a schematic representation of an EPICYCLOIDAL PLANETARY TRANSMISSION WITH A COMMON COMPOUND PLANET WHEEL, having an indirect contact by means of chains or belts between rotary wheels or gear-wheels, incorporating another family member of the present invention.
FIG.1b and **FIG.1d** are samples design, respectively for a HYPOCYCLOIDAL and an EPICYCLOIDAL PLANETARY TRANSMISSION WITH A COMMON COMPOUND PLANET WHEEL, both having an indirect contact by means of chains or belts between rotary wheels or gear-wheels.

FIG.2a is a schematic representation of an EPICYCLOIDAL PLANET WITH A COMMON COMPOUND PLANET WHEEL, having a direct contact by means of gears, incorporating another family member of the present invention.

FIG.2b is a sample design, for an EPICYCLOIDAL PLANETARY TRANSMISSION WITH A COMMON COMPOUND PLANET WHEEL, having a direct contact by means of gears.

FIG.3a is an operating mode depicting the operating characteristics of the ENERGY MAKER MACHINE.

FIG.3b is sample design, for the ENERGY MAKER MACHINE.

A ''PLANETARY TRANSMISSION WITH A COMMON COMPOUND PLANET WHEEL'' has in principle five wheels indirectly connected by means of chains with reference to **FIG.1a** and **FIG.1c**, and directly connected by means of gears with reference to **FIG.2a** :

- Input compound sun wheel **(4)**, necessarily rotary.
- One or multiple common compound planet wheels **(2)**, necessarily triple and rotary inside or outside the ring wheel; and can be quadruple in case of two stationary ring wheels.
- Stationary ring wheel **(3)**, can be two in case of quadruple planets and their radii are necessarily equal if the common compound planet wheels are moving outside those two rings.
- Output compound sun wheel **(1)**, necessarily rotary.
- One or multiple Restorer wheels **(5)** to restore the initial rotational speed, necessarily rotary and connected either to the input sun wheel or to the output sun wheel.

Those wheels are all hold together by means of a carrier **(7)** and a central fixed axle **(6)**; the carrier is necessarily rotary but does not play the role of a wheel.

All parts of a ''PLANETARY TRANSMISSION WITH A COMMON COMPOUND PLANET WHEEL'' are necessarily interacting as follows:
For Z_{R1} the number of teeth of the stationary ring wheel;
For (Z_{S11}, Z_{S12}) the two numbers of teeth of the input compound sun wheel;
For (Z_{S21}, Z_{S22}) the two numbers of teeth of the output compound sun wheel;
For (Z_{p1}, Z_{p2}, Z_{p3}) the three numbers of teeth of the common compound planet wheel connected to the input sun wheel via Z_{p1} and Z_{S12}, to the ring wheel via Z_{p2} and Z_{R1}, and to the output sun wheel via Z_{p3} and Z_{S21};
The ratio ''$(Z_{S12}/Z_{p1})/(Z_{S21}/Z_{p3}) \neq 1$'' is a necessity for a ''PLANETARY TRANSMISSION WITH A COMMON COMPOUND PLANET WHEEL'' to generate mechanical energy.

For (V_{S1}, V_p, V_{S2}, V_R, V_r) the respective rotational velocities of the input compound sun wheel, the common compound planet wheel, the output compound sun wheel, the stationary ring wheel and the restorer wheel;
With reference to **FIG.1b**,
$V_p = V_{S1}/[(Z_{p1}/Z_{S12})-(Z_{p2}/Z_{R1})]$;
$V_p = V_{S2}/[(Z_{p3}/Z_{S21})-(Z_{p2}/Z_{R1})]$.
With reference to **FIG.1d and FIG.2b**,
$V_p = V_{S1}/[(Z_{p1}/Z_{S12})+(Z_{p2}/Z_{R1})]$;
$V_p = V_{S2}/[(Z_{p3}/Z_{S21})+(Z_{p2}/Z_{R1})]$.
With reference to **FIG.1b, FIG.1d and FIG.2b**, V_R is stationary and V_r is the speed of the input or output shaft.
With reference to **FIG.1b**,
the ''n'' factor is
''$(V_{S2}/V_{S1})/[(Z_{S12}/Z_{p1})/(Z_{S21}/Z_{p3})]$'' .
With reference to **FIG.1d and FIG.2b**, the ''n'' factor is
''$[(Z_{S21}/Z_{p3})/(Z_{S12}/Z_{p1})]/(V_{S1}/V_{S2})$'' .

This device's overall system configuration auto generates its own mechanical energy, which it then transforms into electrical energy:

- First, by supplying electric energy from a direct current electricity of a battery **(13)** to a control box **(15)** connected to an electric motor via its AC line **(16)** and its hall line **(17)**. The higher efficiency ''like a brushless'' electric motor model is recommended.
The battery is useful only at start up, it can remain connected in order to recharge or just be disconnected using a switch **(14)**.
- Second, by transforming the received electric energy into a mechanical energy; the electric motor **(18)** drives an input shaft **(12)** of the ''PLANETARY TRANSMISSION WITH A COMMON COMPOUND PLANET WHEEL''.
- Third, by multiplying the input torque by ''n'' factor in order to obtain ''n'' times higher output torque, while the rotational speed of the input and output shaft remains the same; the ''PLANETARY TRANSMISSION WITH A COMMON COMPOUND PLANET WHEEL'' increases the received mechanical energy and transmits it via its output shaft **(10)** to an alternator.
It can be one or multiple transmissions connected in series, with the output of the one linked to the input shaft of the other, in order to achieve a higher output torque while the input and output rotational speed still remaining the same. To cover all the efficiency loss, a hypocycloidal system with a factor ''$n \geq 5$'' is recommended.
- Fourth, by transforming the increased mechanical energy back into electric energy; the alternator **(20)** feeds energy back via its AC line **(21)** to an AC-DC regulator/rectifier **(22)**.

With that AC-DC regulator/rectifier connected via its DC line **(24)** to a battery in order to recharge it accordingly to the chosen configuration, and to the electric motor control box in order to maintain the motor rotational speed even when the battery is disconnected, and as well as to an inverter **(23)**.

Accordingly to the whole system configuration, an alternator such as the ''low rpm three phase permanent magnet alternator model'' that is at least three times more powerful than the used brushless electric motor is recommended.

-Fifth, by harvesting the excess of usable and regulated electric energy via the inverter output **(25)**, since the produced electric energy from the alternator is much more than what the electric motor needs to maintain its rotational speed and what the battery needs to recharge.

A control box with a built-in board to control and monitor the status of the electric motor and battery is recommended. An inverter with higher efficiency features and automatic cut against higher demand of more than the rated power of the whole system configuration is recommended.

According to the present invention, the device is industrially intended to self-produce clean, inexhaustible and sustainable energy; usable at zero emission in automotive devices, tractors, locomotives, aerospace devices, naval devices, robotic devices in general, as well as in power plants and electric generators.

(CLAIMS)

1) Device that generates mechanical energy, characterized in that it comprises a ''PLANETARY TRANSMISSION WITH A COMMON COMPOUND PLANET WHEEL'' having in principle five wheels indirectly connected by means of chains or belts with reference to **FIG.1a** and **FIG.1c**, and directly connected by means of gears with reference to **FIG.2a** :
- One input compound sun wheel **(4)**.
- One or more common compound planet wheels **(2)**.
- One or two stationary ring wheels **(3)**.
- One output compound sun wheel **(1)**.
- One or more Restorer wheels **(5)**.

Those wheels are hold together thanks to a rotating carrier **(7)** and a central fixed axle **(6)**.

2) According to the claim **1**, device characterized in that it comprises a ''PLANETARY TRANSMISSION WITH A COMMON COMPOUND PLANET WHEEL'' of which all parts are necessarily interacting as follows:
For Z_{R1} the number of teeth of the stationary ring wheel;
For (Z_{S11} , Z_{S12}) the two numbers of teeth of the input compound sun wheel;
For (Z_{S21} , Z_{S22}) the two numbers of teeth of the output compound sun wheel;
For (Z_{p1} , Z_{p2} , Z_{p3}) the three numbers of teeth of the common compound planet wheel connected to the input sun wheel via Z_{p1} and Z_{S12} , to the ring wheel via Z_{p2} and Z_{R1} , and to the output sun wheel via Z_{p3} and Z_{S21} ;
The ratio '' $(Z_{S12}/Z_{p1})/(Z_{S21}/Z_{p3}) \neq 1$ '' is a necessity for a ''PLANETARY TRANSMISSION WITH A (triple)COMMON COMPOUND PLANET'' to generate mechanical energy.

For (V_{S1}, V_p, V_{S2}, V_R, V_r) the respective rotational velocities of the input compound sun wheel, the common compound planet wheel, the output compound sun wheel, the stationary ring wheel and the restorer wheel;
With reference to **FIG.1b**,
$V_p = V_{S1} / [(Z_{p1}/Z_{S12}) - (Z_{p2}/Z_{R1})]$;
$V_p = V_{S2} / [(Z_{p3}/Z_{S21}) - (Z_{p2}/Z_{R1})]$.
With reference to **FIG.1d and FIG.2b**,
$V_p = V_{S1} / [(Z_{p1}/Z_{S12}) + (Z_{p2}/Z_{R1})]$;
$V_p = V_{S2} / [(Z_{p3}/Z_{S21}) + (Z_{p2}/Z_{R1})]$.
With reference to **FIG.1b, FIG.1d and FIG.2b**,
V_R is stationary and Vr is the speed of the input or output shaft.
With reference to **FIG.1b**,
the ''n'' factor is
'' $(V_{S2}/V_{S1}) / [(Z_{S12}/Z_{p1}) / (Z_{S21}/Z_{p3})]$ '' .
With reference to **FIG.1d and FIG.2b**,
the ''n'' factor is
'' $[(Z_{S21}/Z_{p3}) / (Z_{S12}/Z_{p1})] / (V_{S1}/V_{S2})$ '' .

3) Device that self generates mechanical energy, which it then transforms into electrical energy, characterized in that it comprises an overall system configuration with reference to **FIG.3a** having in principle:
- a battery **(13)** connected to an electric motor via a control box **(15)**.
- An electric motor **(18)** connected to the input shaft **(12)** of a ''PLANETARY TRANSMISSION WITH A COMMON COMPOUND PLANET WHEEL''.
- A ''PLANETARY TRANSMISSION WITH A COMMON COMPOUND PLANET WHEEL'' **(19)** connected via its output shaft **(10)** to the alternator.
- An alternator **(20)** connected via an AC-DC regulator/rectifier **(22)** back to the battery, to the electric motor control box and to the inverter.
- An inverter **(23)**.

(DRAWING)

FIG. 2a

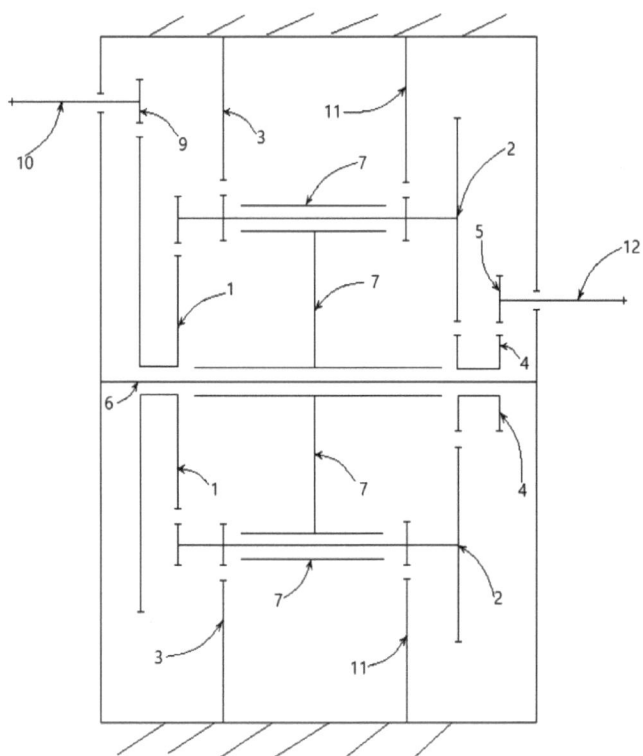

FIG. 2b

SAMPLE DESIGN "n" Factor ≈ 1.5		Compound input sun gear wheel (4)		Common compound planet gear wheel (2)			
FIG.2a	"Z" Number of teeth	Z_{S11}	Z_{S12}	Z_{p1}	Z_{p2}	Z_{p3}	Z_{p4}
		8	8	24	8	8	8
	"V" Rotation speed	V_{S1}		V_P			
		! V_{S1}		$V_{S1} \div 3.2$			

Ring gear wheel (3),(11)		Compound output sun gear wheel (1)		Output compound restorer gear wheel (9)		Input compound restorer gear wheel (5)	
Z_{R1}	Z_{R2}	Z_{S21}	Z_{S22}	Z_{r11}	Z_{r12}	Z_{r21}	Z_{r22}
40	40	24	48	8	Output shaft (10)	8	Input shaft (12)
V_R		V_{S2}		V_{r1}		V_{r2}	
stationary		$V_{S1} \div 6$		$\approx V_{S1}$		V_{S1}	

FIG. 1a

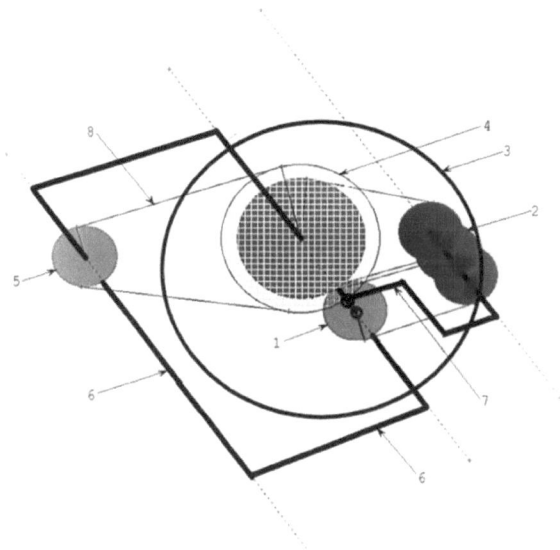

FIG. 1b

SAMPLE DESIGN "n" Factor ≈ 2.03		Compound input sun gear wheel (4)		Common compound planet gear wheel (2)			
FIG.1a	"Z" Number of teeth Model No.:428	Z_{S11}	Z_{S12}	Z_{p1}	Z_{p2}	Z_{p3}	Z_{p4}
		98	45	13	14	14	–
	"V" Rotational speed	V_{S1}		V_P			
		$1 \cdot V_{S1}$		$8.78 \times V_{S1}$			

Ring gear wheel (3)		Compound output sun gear wheel (1)		Compound restorer gear wheel (5)	
Z_{R1}	Z_{R2}	Z_{S21}	Z_{S22}	Z_{r1}	Z_{r2}
80	–	16	Output shaft	16	Input shaft
V_R		V_{S2}		V_r	
stationary		6.14 x V_{S1}		≈ 6.14 x V_{S1}	

FIG. 1c

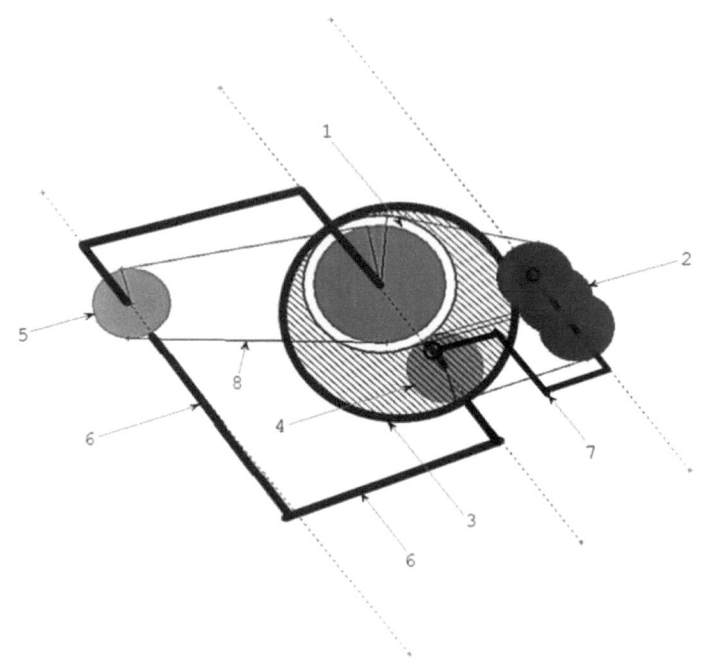

FIG. 1d

SAMPLE DESIGN "n" Factor ≈ 1.182			Compound input sun gear wheel (4)		Common compound planet gear wheel (2)			
FIG.1c	"z" Number of teeth		Z_{S11}	Z_{S12}	Z_{p1}	Z_{p2}	Z_{p3}	Z_{p4}
	Model No.:428		Input shaft	16	16	16	16	–
	"v" Rotational speed		V_{S1}		V_P			
			! V_{S1}		$V_{S1}/1.5714$			

Ring gear wheel (3)		Compound output sun gear wheel (1)		Compound restorer gear wheel (5)	
Z_{R1}	Z_{R2}	Z_{S21}	Z_{S22}	Z_{r1}	Z_{r2}
28	-	24	20	16	Output shaft
V_R		V_{S2}		V_r	
stationary		$V_{S1}/1.2692$		$\approx V_{S1}$	

FIG. 3a

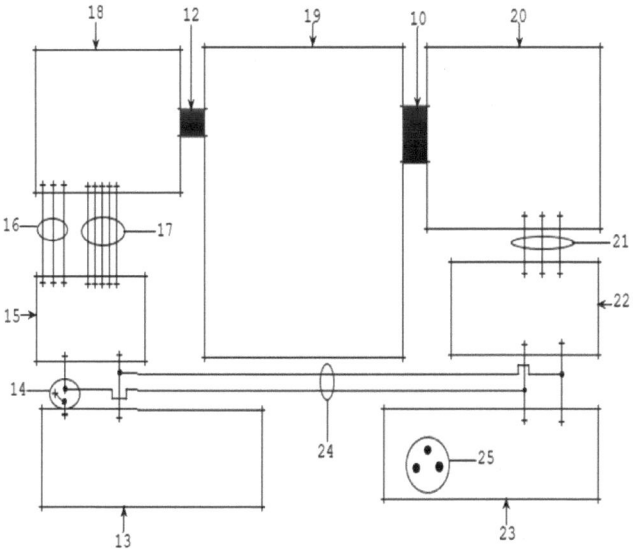

FIG. 3b

SAMPLE DESIGN		Lithium Battery (13)	Electric motor control box (15)	Blushless electric motor (3)
FIG.3a	specifications	36V 10AH	36V 500W	36V 500W
	System efficiency	≥ 80%		
	System rated power		Input/Feedback	output
			500W	≥ 400W
Generated power		Inverter (23)		
		specifications	DC 36V-48v to AC 220 v 600 W	
		efficiency	≥ 85%	
		Input power	620W	
		Output power	≈ 500W	

Planetary transmission with common compound planet ''n'' Factor **(19)**		Permanent magnet Alternator **(20)**		AC-DC regulator/rectifier **(22)**	
≥ 5		3 phase AC 48v 1500W		AC 48V - DC 48v 1500W	
≥ 80%		≥ 70%			
Input	output	input	output		
≥ 400W	≥ 1600W	≥ 1600W	≥ 1120W	feedback	output
				500W	620W
Generated power		Inverter **(23)**			
		specifications		DC 36V-48v to AC 220 v 600 W	
		efficiency		≥ 85%	
		Input power		620W	
		Output power		≈ 500W	

HAKIZA 1 (THE ENERGY MAKER MACHINE)

1st HAKIZA 1

Details:
This Prototype aims to provide material evidence to the theory of universal mutation highlighting the bond (MICRO-INVISIBLE-MACRO),
it is a machine that provides an industrially crucial and revolutionary solution in terms of energy (ANY GIVEN ECONOMY IS JUST A TRANSFORMED ENERGY) and climate:

. 1st artificial machine to produce perpetual motion thanks to dark gravity
. 1st artificial machine to generate mutational energy (invisible - visible)

According to the present invention, the device is industrially intended to self-produce clean, inexhaustible and sustainable energy; usable at almost zero emission in automotive devices, tractors, locomotives, aerospace devices, naval devices, robotic devices in general, as well as in power plants and electric generators.

I PERSONALLY URGE THE WORLDWIDE YOUTH TO EMBRACE THE NEW-ERA AND REINVENT EVERYTHING INSTEAD OF KEEPING ON SCREAMING FOR HELP.

More than two millennia ago, ARCHIMEDES said:
« *Give me a place to stand and with a lever, I will move the whole world* ».

I, Olivier HAKIZIMANA (3rd son of Onésime HAKIZIMANA and Régine NIBIZI, born in KINAMA/BUJUMBURA/BURUNDI on April 14, 1995), I quote:

« *Here is a HAKIZA and its invisible fulcrum, let's preserve the world* ».

* TO COUNTER THE CHALLENGES FACING THE ENTIRE WORLD, I ALLOW ANY INDIVIDUAL OR COMMUNITY NON-PROFIT CORPORATION TO EXPLOIT MY INVENTION.

Be careful ! "Within the Oneness of the Cosmos, one is different and can only become the best or the worst version of oneself".
So, I defy anyone who would cheat or use my invention to negatively accelerate the fate of humanity in favor of his own ideologist and illusionist supremacy.

** I HOPE EVERYONE WILL MAKE GOOD USE OF IT … OTHERWISE WILL BE UNCONDITIONALLY AND MERCILESSLY LAUNCHED:

The HAKIZA 2 (**THE JUSTICE MAKER**)
Details: Top-secret.

Many thanks to members of the National Defense Army Corps of BURUNDI; the others thought I was a young arrogant madman while you encouraged me.

Glory to BURUNDI, glory to AFRICA, and justice throughout the world.

*** I MIGHT BE CRAZY, BUT I'M SURE THERE IS A WAY TO IMPROVE THE LIFE EXPECTANCY BY HACKING BIOLOGICAL FUNCTIONS THAT ALLOW LIFE:

The HAKIZA 3 (**THE NANO-HAKIZA**)
Details: Still under development.

Note:

During the game,
curious are those who take risk;
anxious are those who become the risk itself;
victorious are those who understand it and act on time , to stay in the game .

Pendant le jeu,
curieux sont ceux qui prennent le risque ;
anxieux sont ceux qui deviennent le risque en-soi ;
victorieux sont ceux qui le comprennent et agissent à temps , pour rester dans le jeu .

END OF THE SUPREMACY WAR = BEGINNING OF THE PRESERVATORY WAR.

CONCLUSION

3. The Introduction to the theory of everything

To conclude, based on facts observed during the operation of the first artificial machine "The HAKIZA 1", which (through the phenomenon of mutational frictions) is able to mutate energy from the invisible to the visible (vice versa); it ends up comparing and explaining the strange behaviors of orbital systems in the entire universe .

Seeing that we share the same conviction as Einstein and Bohr (despite their biggest mistake of denying each other's theory when they were both right, seeing this theory); a theory that you are not able to explain to six-year-olds, you yourself have not understood it.
That is why this doctrine of truth is conceptually one of the simplest with just one fundamentally-mathematical generalized equation... after all; this is an introduction to the unification of sciences.

A. The natural HAKIZA

(Mystery of the so-called ''dark gravity, black or white holes and visible spirals around'')

Out of simple curiosity, anyone who analyzes well the movement of a circle that rolls around another stationary circle must think like **LAPLACE** and **BOHR** of spirographs generating cycloids (epicycloids or hypocycloids).
The latter give a prediction of the almost exact trajectories of the movement of the visible whole.

With reference to **figure 18**

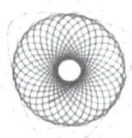 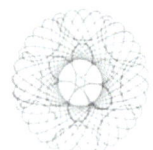 

- According to **LAPLACE** (celestial mechanics: ''the moon describes an almost circular orb around the earth and one would wrongly assume that the moon does not rotate; but when it is observed from the sun, it describes epicycloids having centers on the terrestrial orb's circumference... and so forth'').

- According to **BOHR** (the planetary model of the atom: ''some theoreticians might be questioning this classic model of the atom by assuming that a spin of particles is the so-called intrinsic-property since one cannot determine the axes of rotation, but still the velocity of the electron particularly in the 1st Bohr orbit of hydrogen **v= c/137** is correct with 1/137 being the fine-structure constant appearing almost everywhere; so apart from having the so-said internal angular momentum, the electron also has an orbital angular momentum and unless it is proved to be wrong which would turn out to be a contradiction to the existence of nature itself... cease to play the quantum- mechanics- fundamentalist more than the quantum-theory-pioneer Bohr... otherwise you will keep ending up either with theologies considering one-same electron everywhere or farther enigmatic issues like the proton spin crisis since the complex motion of the nucleus is electromagnetically linked to the complex motion of the orbiting-electrons'').

-To complete them according to HAKIZIMANA (hypothesised paradoxes of a visible body in motion : '' regardless of the mass and no matter how the axes of rotation might evolve, as long as a given body represented as a three-dimensional sphere analogue to a two-dimensional circle and a one-dimensional point is visible and is moving in a given orbit either by rolling-sliding along the orbit due to any given force at a human being scale, or by rolling-sliding in a free-fall along the orbit due to a gravitational force at a large scale, and or by completely sliding in a free-fall all along the orbit due to an electromagnetic force at a small scale, after achieving a complete revolution, that body always ends up with an excess or a missing of one turn; this is extrinsic and that is how both particles and stars end up having a rotational motion... in other words, everything rotates, it is all about fundamentally and dynamically orbital systems in which a perpetual motion can be maintained thanks to the mutation between the invisible and the visible'').

-Moreover by paying a particularly mathematical (according to a study of spirals by **GALILEO**, or **TORRICELI**, or **BERNOUILLI**, and others…) attention to the equation of a mutation, within a homogenous and isotropic universe with hydrogen as the most abundant element, one ends up noticing that the basic mutational factor ''n'' is almost the same everywhere; and consequently the equation of a mutation $X_m = X \cdot (n_1 \cdot n_2 \ldots n_p)$ can become $X_m = X \cdot n^p$ with ''n'', ''p'' and ''X'' being strictly non-zero positive.

Since "n" is defined as the basic mutational factor which by mutational singularity belongs in the range $]0, 1[\cup]1, 2[$ hence "$n \neq 1$".

"p" is defined as the number of ingoma growing arithmetically (such as a polar angle of a logarithmic spiral) within a HAKIZA.

"X" is defined as a given physical quantity, needed for the triggering of the movement that must therefore be positive and greater than zero (such as a radius vector of a logarithmic spiral for a zero polar angle).

And "X_m" is defined as a mutated physical quantity growing in a geometric way (such as a radius vector of a logarithmic spiral).

Finally... the equation of a mutation that can become an exponential function "$X_m = X \cdot n^p$", it is also likely to be extremely able to become the polar equation of a logarithmic spiral admitting a symmetry "$X_m = X \cdot (1/n)^p$".
It is so much likely to be a kind of golden spiral that frequently governs the structure of the visible and/or observable universe; accompanied by the certain total collapse of NOETHER's theorem (since the symmetry is not leading to the principle of energy and/or momentum conservation).

Furthermore, an ingoma plays exactly a black-white twin hole (wormhole) ; with physical quantities both emerging strictly from the invisible (on one side of the transmission, in real time commonly called past) and disappearing strictly towards the invisible (on the other side of the transmission, within an imaginary time commonly called both past and future).
A total collapse of the principle of causality and entropy (the future is influencing the past, vice versa).

- Hence, we allow ourselves to redefine a natural HAKIZA as the ideal staggered representation of the fundamental structure of the visible depicting dark gravity, by allowing the mutation between the invisible and the visible, with the number and the configuration of natural ingoma being the main characteristics. The fundamental interactions (respectively gravitational and electrical) have to be rewritten mathematically as follows:

$F_m = (G \cdot m_1 \cdot m_2 / R^2) \cdot n$
and its symmetry $F_m = (G \cdot m_1 \cdot m_2 / R^2) \cdot (1/n)$

$F_m = [(K \cdot |q_1| \cdot |q_2|) / R^2] \cdot n$
and its symmetry $F_m = [(K \cdot |q_1| \cdot |q_2|) / R^2] \cdot (1/n)$

G is the **NEWTON**'s gravitational constant; m_1 and m_2 are masses; q_1 and q_2 are electrical charges; R is the radius or distance between the centers of two masses or charges in question; **n is the basic mutational factor of HAKIZIMANA**; K is the **COULOMB** constant.

F_m is a mutated or mutational force that counterbalances gravitational braking (at the micro or macro scale-level, it is at once both repulsive in case of centrifugal ingoma and attractive in case of centripetal ingoma) in order to allow integrity of natural orbital systems and perpetual movement of the visible whole (for any natural dynamically-orbital system, the determination of a mutational factor requires the use of different advanced mathematical skills and methods not mentioned in this introduction for the theory of everything).

The HAKIZA-1 is finally a direct proof to the Yang–Mills existence and mass gap, a birth of a new theory unifying the infinitely small and the infinitely large; that we name:

`''THE UNIVERSAL MUTATION''`.

BRAINSTORMING : the scientific story that opposed BOHR and EINSTEIN or their supporters is indeed comparable to this story below:

   An animal (so big enough, black on the left-side and white on the right-side) passed in a village between two hills. Because of the vastness of the size of this animal's body, no one was able to see both sides of the beast at once. It was after its passage that the villagers of those two hills gathered to discuss what they had just observed.

Unfortunately, they never agreed on the colour of the animal in question and... Yet they still ask themselves the same question: "black... or... white?" ... **Fortunately, in Kirundi we say: "umugani ugana akariho".** The animal always comes back and what was just a story becomes a reality.

This is it... here we are... Blinded by the immensity of the cosmos, "the only mistake that the two men or their supporters committed" was to themselves limit by the supposed impossibility of exceeding the speed of light "the celerity" of which they all believed untouchable.

Consequently, they all became unable to surpass or break through this immensity of the animal, so that they could first understand the invisible in order to be able to observe the colour on the other side, and as a result:

→ **EINSTEIN** is unable to explain to six-year-olds his theory on the hidden variable that really exists and ends up looking implausibly towards quantum theory that really exists too, but of which we cannot explain to a six-year-old child either (according to **FEYMAN** : "If you think you understand quantum mechanics, you don't understand quantum mechanics").
And their supporters who over the years will create the theories (strings, quantum gravitation, MONDs, WIMP...), completely insufficient (with zero empirical and material evidence) from a scientific point of view, or just fundamentally wrong speculations from my personal point of view.

 Ultimately, most theories emanate generally from the two fundamental interactions (gravitational and electrical), which means that some of them are fundamentally correct but must be re-examined...
for instance, we consider the equations of **DIRAC** and **EISTEIN** (without cosmological constant = desperate act) as a testament of creation... but we so far consider the equation of universal mutation as creation itself.

Here is the major breakthrough of the present theory: " information does not need to reach the speed of light to overtake light;
it is all about being mathematically within the imaginary time; which is the equivalent of being physically in a dynamically-orbital system".

And make no mistake about this: "Nobel-prize of physics 2022 proves and assumes the inexistence of hidden variable (local); the present theory-introduction proposes a hidden variable (no-local)".

3.1 The theory of everything via Mutational factorization

From the conceptual point of view, by starting to compare **HAKIZIMANA**'s paradox to that (among others: **YOUNG**'s double slit, **SCHRÖDINGER's** cat thought experiment, **BOHR-EINSTEIN** photon-box experiment), we notice one of the most important simultaneities between the infinitely small and the infinitely large:

"the duality behavior appealing to the singularity notion".

Admitting that the invisible (both converging towards the micro and diverging towards the macro) is everywhere in the visible, that they are linked by frictions in each of the points of the visible whole, and that they interact specifically via a phenomenon of mutational frictions (obeying the three mutational laws); is to admit that these three laws are applicable at the scale-level of the infinitely small as well as at the scale-level of the infinitely large (the operating mode of the HAKIZA 1 is a living evidence).

We deduce that the common behavior of duality (*invisible-visible or wave-corpuscle, which embodies the principle of superposition of states*) is due to a connection with an infinitely high speed of the invisible to the visible (MICRO-INVISIBLE-MACRO), finally allowing us to confirm with certainty the constantly simultaneous existence of the quantum entanglement and hidden variable (no-local).

In other words, within a fundamentally-dynamically-orbital universe, there is a common moving invisible thread (can be dubbed the so-said dark matter) everywhere (at any scale); whose mutated or mutational speed can infinitely increase to overtake the speed of light (celerity) according to a mutational factor (can so far be dubbed the so-called hidden "no-local" variable) depending on characteristics and conditions (the information within imaginary time = the information within a dynamically-orbital system) of an ingoma within a HAKIZA.

According to the first mutational law, this mutated velocity $C_{mutated}$ (once raised to infinity, allows quantum entanglement) is mathematically determined as follows:

$C_m = C \cdot N$ with a symmetry $C_m = C \cdot (1/N)$

IN GENERALLY SPEAKING ENERGY EQUALS INFORMATION:

$E_m / E = N$ with a symmetry $E_m / E = (1/N)$

$$\rightarrow \quad E_D = E_m - E$$

''$C_m > C$'' is a mutated or mutational speed of light (*the speed of information within imaginary time*).

''C'' is the speed of light (*the maximum speed of information in real time*).

''N'' is the total mutational factor (*the non-local hidden variable = information*) within a natural HAKIZA (*dynamically-orbital systems = the so-said black-white twin holes = wormholes*).

''E_m'' is the mutated or mutational energy (*disobeying the principle of conservation*) that includes and explains the accelerated expansion of the invisible space metric of the visible universe (*dark gravity*).

''E and E_D'' are respectively mechanical energy (*obeying the principle of conservation*) and invisible energy (*the so-said dark energy and/or vacuum energy*).

MORAL: Finally, within a fundamentally-dynamically-orbital universe, the phenomenon of mutational frictions (regulating the interaction between the invisible and the visible) occurs or can therefore occur at any given point of the visible and/or observable whole.

Besides, I am a small visible being composed by lot of dynamically orbital systems at the atomic scale; then I'm an energy maker, a visible universe with an infinite number of "twin black-white holes (wormholes) = barycenter of the visible and the invisible".

what a joke...we were/are seeking our own self-identity... any well-thinking biological being can play the role of a so-said **"Santa Claus"** whose no wish can escape as long as there is a will from kind children endowed with the necessary energy to trigger the mutation of their vows from the invisible to the visible !!!... .

We conclude that life is none other than the result of a universal mutation that the visible universe unceasingly undergoes, when the invisible interacts with the visible (micro-invisible-macro) via mutational frictions, regardless of the referential frame and dimension.

Here are the crazies playing hide and seek... the confused eldest having fun with the confusing 9-year-old and the imaginative 6-year-old:

The eldest, facing a wall, launches:
"Are you hidden?... I'm coming to find you".

The 9 years old kid replies:
"You can't, we are everywhere and nowhere".

The lost elder, exclaims with a laugh:
"What! ... No bro... God does not play dice"!

The 9 years old kid asks with a laugh:
"Who are you to tell God what to do"?

Instead of the lost elder, the 6 year old kid responds without hesitation:
"Santa Claus".

3.2 Universal thinking via mutational invariance

The visible human being infinitely small with respect to the immensity of the visible whole, is shaped by his faculty to invisibly and infinitely think... which makes him doubtful (primarily according to black Africa) or platonic (according to the immoral whitewashing of the power of knowledge) regardless of his level and ability to understand the cosmos.

Moreover, at this very moment, writing down this theory, I say to myself:
I have just clarified one of humanity's most hidden truths (the creation mechanism).
But then... why me and now? Why even ask ... if consciousness evolves according to morality and that morality can be anything as long as we think we are right?
One could even end up deluding oneself and think that the end justifies the means... nevertheless, should we not first ask ourselves what justifies the end ... why even the end, the beginning?

And by a simple logic that the mutational invariance law proves to us:
the visible reality and the invisible reality turn out to be almost the same in any dynamically orbital system where the movement of the infinitely small is shaped by the infinitely large.
In short, we turn around in circles; the epistemologically objective is the fruit of the ontologically subjective whose well-thinking biological beings are both the bearers and the followers.

I conclude that a human being had ever been, is not and will never be able to prevent himself from wondering ''the why ... '' and will always end up answering ''the how...'' as a limit point; in other words:

THERE IS A HOW FOR **"EVERYTHING EXCEPT THE-WHY-ITSELF"**.

Amosozi y'imbeba uyabonera mu kamashu!

Note: "Curiously, more we go around in circles, more we surprisingly understand the trap; keep going".

3.3 Universal Certainty via Mutational singularity

Scientifically, everything is interconnected and ''no imagination'' means ''no sciences''.
Therefore, it is antiscientific to separate ''the imagination of the Beyond (Supreme Being beyond its own definition, present in everything and beyond everything)'' from ''the whole (invisible + visible)''.

Indeed, it is obvious that according to the operating mode of the creation mechanism (the HAKIZA 1 is a living proof), only the Beyond of that mechanism ''accordingly to his will'' can design, trigger, retrigger and/or end the movement of the latter; result :

There is a creation of the fine-structured whole ... then we experience the mutation of the structure of the whole ...and finally there will be a disappearance of the whole and/or its structure.

And by a simple logic that the mutational singularity law proves to us:
during the triggering, or re-triggering and/or ending of a mutation, depending on the will of the trigger, the latter can make the whole appear or disappear except the whole that is equal to himself (''his own universal will that determines the information about the conditions of a natural HAKIZA preceding the PLANCK's time and radius''= THE-WHY-ITSELF= THE-EXISTENCE-ITSELF).

Finally... it is neither zero or less nor two or more, it is a singularly and a necessarily existing indefinable infinity; in other words:

"IMANA= THE BEYOND = THE ALL-POWEFULL INVISIBLE ETERNAL ONE GOD".

Vichwa ngumu daima kuwepo! Shauli yenu.

Note: "Everything is possible except the existence of a second God = the nothing = the impossible existence".

HAKIZIMANA

THE UNIFICATION OF THE VISIBLE WHOLE VIA THE INVISIBLE INVARIABLY AND SINGULARLY MUTATIONAL

Breakthrough: "information does not need to reach the speed of light to overtake light; it is all about being mathematically within the imaginary time, which is the equivalent of being physically in a dynamically orbital system".

Within any dynamically orbital system, the visible whole is created and lost in the invisible; it is just a mutation of which only the trigger of the movement decides its beginning, its magnitude and its end.

And life is none other than the result of a universal mutation that the visible universe constantly undergoes, when the invisible interacts with the visible via mutational frictions, regardless of referential frames and dimensions; with a cycle "(invisible = information within imaginary time) $\leftrightarrows$ (visible = information within real time)" depending on the universal will of which everything is possible except "an existence that is equal to it" :

« THE IMPOSSIBLE EXISTENCE = THE NOTHING = THE SECOND GOD ».

www.ingramcontent.com/pod-product-compliance
Lightning Source LLC
Chambersburg PA
CBHW040056250526
45473CB00042B/2465